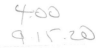

ALL NEW

BATHROOM
IDEAS THAT WORK

ALL NEW BATHROOM
IDEAS THAT WORK

DAVID SCHIFF

The Taunton Press

The Taunton Press
Inspiration for hands-on living®

The Taunton Press, Inc.,
63 South Main Street, PO Box 5506
Newtown, CT 06470-5506
Email: tp@taunton.com

Editor: Carolyn Mandarano
Copy Editor: Nina Rynd Whitnah
Indexer: Jim Curtis
Interior design: Carol Petro
Layout: Sandra Mahlstedt
Illustration: Mario Ferro
Cover Photographers: Front cover: Eric Roth; Eric Roth, design: Butz+Klug Architects, Boston; courtesy Brizo® Kitchen and Bath Co.;
photography by Troy Theis, design: Martha O'Hara Interiors (top, left to right); Brian Vanden Brink, design: Breese Architects (bottom)
Back cover: Eric Roth, design: Thomas Buckborough Design, Boston; Eric Roth; Hulya Kobalas, design: Michele Hogue Interior Design (top,
left to right); Brian Vanden Brink, design: Polhemus Savery DaSilva Architects Builders (center); courtesy Kohler Co. (bottom left); Brian Vanden
Brink, design: Hutker Architects (bottom right)

Library of Congress Cataloging-in-Publication Data

Names: Schiff, David, 1955- author.
Title: All new bathroom ideas that work / David Schiff.
Description: Newtown, CT : Taunton Press, Inc., [2018] | Includes index.
Identifiers: LCCN 2018003938 | ISBN 9781631868788
Subjects: LCSH: Bathrooms–Remodeling. | Bathrooms–Design and construction.
Classification: LCC TH4816.3.B37 S35 2018 | DDC 690/.42–dc23
LC record available at https://lccn.loc.gov/2018003938

Printed in the United States of America
10 9 8 7 6 5 4 3 2 1

ACKNOWLEDGMENTS

This book is an amalgam of the ideas of many architects, designers, and other professionals who took the time to share their knowledge and projects with me. Because these pros work in the field each day, they helped ensure that this book contains wisdom that is not only comprehensive but also on the cutting edge of bathroom design. A heartfelt thanks to architects Alex Bertraun, Paul Hannan, Mark Hanson, Jeff Krieger, Sophie Piesse, Anni Tilt, Marc Sloot, Mike Waters, and Deborah Woodward; bath designers David Nault and Toni Sabatino; lighting designer Peter Romaniello; and universal design experts Richard Duncan, Eric Listou, and John O'Meara.

Thanks to photographer Randy O'Rourke, who traipsed around the country photographing bathrooms for this book, and to all the other photographers who contributed their work. I'd also like to thank all the homeowners who were kind enough to welcome Randy into their homes.

At Taunton Press, thanks to editors Peter Chapman and Carolyn Mandarano, who helped shape the text. Carolyn, it was a pleasure to work with you again after all these years. And thanks to the design and production staffers who always do an amazing job of melding words and images into books that are a pleasure to read.

CONTENTS

INTRODUCTION

Today's housing market is a tight one, as my daughter Aliza discovered this year when she and her husband Chris sought to move from a rental in Arlington, Virginia, to their own home in Warwick, New York. Several times they packed the kids into the car to make the 265-mile trip to look at a house. Each time, they got to Warwick too late—someone else had made an offer. They finally snagged a house by making an offer the day it went on the market. They had seen the house only in a video walk-through on their real estate agent's cell phone.

Data from Realtor.com shows that Aliza and Chris experienced a nationwide trend: that houses are selling rapidly because inventory is in short supply. They also report that "listing prices in the residential real estate market remain near historical highs."

As a result of the short supply of homes, many people are choosing to remodel or add to their current homes, notes Metrostudy, a market research firm owned by Hanley Wood, the parent company of *Remodeling* magazine. And of all the projects homeowners undertake, 26% are bathrooms, second in frequency only to kitchens at 31%, according to a 2017 survey by the Houzz® home design website.

Think of your bathroom project as an investment in improving your family's lifestyle. As you'll discover in this book, there has been an explosion in choices for fixtures, finishes, and even lighting. Incorporating universal design features, green building, and products that tread more lightly on the earth's resources is growing as well, and you'll find examples of those options in these pages, too.

It'll be a challenge to sort through it all, but fun, too. Time spent doing your homework and working closely with any pros you hire will help you create a comfortable and safe environment you'll enjoy for years to come.

But don't make the mistake of thinking of the bathroom as a financial investment. Although bathrooms are a popular project, you're not likely to recoup your remodeling dollars at resale time. Mid-range bathroom remodels cost an average of $12,000 and returned only 65% of that, according to the 2017 cost vs. value study published by *Remodeling*. Bathroom additions, at an average cost of $43,000, recouped only 54%—the worst return among projects.

There are a few reasons for the low return: Complex projects in general return less than simple ones, and exterior projects return more than interior projects. Plus, a prospective buyer might love your house but have completely different tastes when it comes to bathroom fixtures.

As you read on, you'll discover that there's a lot to consider in designing today's bathroom. I hope you enjoy the journey as you plan and build the room that perfectly fits your needs and style.

PLANNING

Start with an assessment of your family's needs, desires, and tastes,

then work with the right pros

YOUR

to create a bathroom that you'll enjoy

for many years to come.

BATHROOM

Designing a bathroom is an exciting and creative project with more options than ever before. Today's bathrooms have evolved away from the plain little room housing a sink, tub, and toilet. In general, we expect bathrooms to be larger, with more functions and much more style.

Today, you'll find bathrooms with vanities that are more like fine furniture pieces, chandeliers, built-in entertainment centers, even fireplaces. Today's master bath typically has two sinks—either a dual-sink vanity or perhaps two completely separate sink areas.

Increasingly, the familiar tub/shower combination is being replaced by a deeper separate soaker or spa tub as well as a dedicated shower that often features a seat, multiple showerheads, and its own storage. Many master baths have no tubs at all—it's common to dedicate the space to a more luxurious shower, especially if the family bathroom has a tub.

Even a relatively modest bathroom makeover, say, replacing fixtures without changing the layout, presents more choices then ever before. Showerheads alone come in seemingly endless permutations. Beyond choosing a finish, there are heads that massage, heads that mimic rain, heads that can be adjusted for height, or heads that are handheld. You can buy a chromotherapy showerhead—one with a light that changes color to suit your mood. Want music in the shower? A showerhead with a built-in Bluetooth® wireless speaker may be the ticket.

While all of these choices let you customize your bathroom remodel to your needs and desires, they can also be a bit bewildering. The more extensive the project, the more it makes sense to hire a professional designer to help you sort through it all. This chapter will help you decide if you need a pro designer and will explain the roles of the other pros you may hire. I'll help you organize your thoughts about what you like and how you'll use the bathroom, both now and also as your needs change in the future. This will prepare you to communicate effectively with a designer and/or your contractor to make sure your bathroom dreams are realized.

The trend toward larger showers is an invitation to get creative. This shower incorporates stone tiles of various colors and shapes. It also has niches, a seat, a handheld showerhead, and body spray nozzles at various heights.

Here, the contemporary idea of a large shower and a separate soaking tub is blended beautifully into a classic home.

A master bath will often have two sinks. Vanities that look like freestanding furniture, like this one, are quite popular.

The tub is likely to be separate from the shower these days. What could be more relaxing than a soaking tub with a beautiful view?

Defining Purpose

The first step in narrowing your choices is thinking about who will use the room and what they will do there. Is it a downstairs powder room that will be used by guests? In addition to the obvious toilet and sink, that calls for a mirror that's well lit for adjusting hair and makeup. It might be the perfect spot for a stylish flourish, such as an unusual vessel sink.

If you're creating a master bath for two, how and when will you both use it? If you rush off to work at about the same time, two sinks will be a top priority, and you may want to put the toilet behind its own door if possible. Do either of you enjoy long soaks in a tub or would you rather devote the space to a bigger more luxurious shower? Depending on the answer, you might consider a separate tub and shower, if you have the space.

Will kids use the bathroom? In general, babies and little kids need a bathtub, although there are ways around that. And if they are going to need a step up to reach the counter, stools that fold into the vanity kick are a terrific way to get them out of the way. Remember, too, that kids are messy and splash a lot. You'll want to use tough, waterproof materials such as tile and solid surface countertops—save wood counters and vessel sinks for the powder room. And you'll want to design for easy cleaning—for example, undermount and integral sinks eliminate the grime-collecting seam around a top-mounted sink.

top right · This bathroom is used by two boys. Kids can be messy, and this trough-style sink with two faucets is less work to keep clean than two separate sinks.

right · Putting the toilet behind a door within the bathroom is a great idea for a shared master bath. Here, a pocket door slides into the wall so it doesn't get in the way as a swinging door would.

left • This ultra-modern bathroom features a rectangular tub, a translucent toilet enclosure, and a large shower with a ceiling-mounted rain showerhead.

below • Powder rooms get relatively light use, making them the perfect place to get fancy by including elements like this old-fashioned-looking washstand with a delicate vessel sink. The sconces on both sides of the mirror look elegant and are perfectly positioned to light the face when checking makeup.

If you are remodeling, give some thought to what you don't like about your old bathroom. As you take your first flip through this book you'll notice features called "Putting It All Together." Many of these focus on how architects solved problems in existing bathrooms. For example, if your bathroom is too dark, be sure to read "An Un-Bathroom Vibe" on p. 116 and "Opening Up a Dark, Narrow Space" on p. 170 to see how architects greatly increased natural light without increasing the size or number of windows. Is the bath too small? Even if you can't increase the square footage, there are strategies for using the space more efficiently and even visual tricks you can employ to make the space *seem* bigger. Check out "Small but Airy Casita Bath" on p. 18, "A Wet Room Optimizes Available Space" on p. 70, and "Angles Make a Small Bath Seem Spacious on p. 54.

Don't Move the Fixtures

Installing water pipes is expensive, especially in an existing house. So if you are remodeling an existing bathroom, you'll save lots of money by placing the new toilet, sink, tub, and/or shower right where the old ones were. If you are creating a bathroom in a new house, you'll have some flexibility about where to place the tub, shower, sink, and toilet, although even then economics can affect your decision. For example, if one wall of an upstairs bathroom were directly above the kitchen sink wall, placing a water supply—either the bathroom sink or the tub and/or shower faucet—on that wall would be a cost-effective choice.

The bold geometry of this tile floor is made all the more dramatic by the white walls, ceiling, and tub.

top left · A large archtop window provides plenty of daylight and a great view for the platform tub in this classically elegant bathroom.

top right · This bathroom celebrates the warmth of wood. The all-glass shower enclosure virtually disappears, while a delicate glass chandelier adds a touch of whimsy. Café-style curtains cover the lower part of the window, providing privacy for this second-story room without blocking the light.

left · Wood is not typically used to make bathtubs, but with today's waterproof finishes and proper maintenance, it's certainly a feasible choice. And no one would deny its beauty.

NOW FANTASIZE!

Reality will come into play soon enough, but start by finding stuff you really like. Go ahead and buy a bunch of kitchen and bath magazines and tear out pages that appeal to you, perhaps just because you like the style of a bathroom or a faucet, or you discover a showerhead with a function that appeals to you. Paste the pages into a scrapbook. Flip through the photos in this book, too, and put a sticky on those you like. You can lose yourself for hours on the Internet just by Googling "bath design." Online sites like Houzz can help you identify styles and items you like, and they also identify the designers and architects. At this point, just have fun. The exercise will help you define your taste and desires and, along with actually reading this book, will make you aware of the latest options. Flip through your scrapbook while discussing ideas for the bathroom with your family—some may want to make their own scrapbook.

This process will inform your decisions when you do get down to the hard work of determining what you really need, what you really want, and what trade-offs you'll make to accommodate the space and budget you have to work with.

top right • Most people don't have the space or a budget for a bathroom of this size, but it can still provide great design ideas, such as the LED strip lighting in the ceiling soffit and the use of tile of various sizes and colors to define areas.

right • It doesn't necessarily take elaborate and expensive fixtures to make a bathroom into a relaxing retreat. Here, an old-fashioned claw-foot tub and a comfy chair do the job just fine.

Randomly placed brightly colored tiles create a bold and cheerful shower enclosure.

Choosing Professionals

For a simple bathroom redo—perhaps new fixtures in the same positions, new tile, and a coat of paint--you might only need to hire a general contractor. As the job gets more complex, you'll want to consider other pros, including a bath designer, an architect, and perhaps even a lighting designer.

Whether you are hiring just a contractor or bringing in other pros, it's important that they are people with whom you have a rapport—folks who you sense are listening to your ideas. They certainly should have ideas of their own, and you will want to benefit from their experience. But at the end of the day, it's *your* vision they are hired to execute, not what they may think the bathroom should be.

If you do hire more than one pro, think of it as putting together a team, with you as the leader. Not only do you need to work with everyone on the team, but it also is equally important that everyone on the team is willing and able to work together. You may need to interview several pros in each field to find the right one. Ask pros for portfolios of their work and references for previous clients you can call.

Here's a look at the roles and responsibilities of the pros you might hire.

Architect. If your bathroom involves major structural changes such as moving walls, or if the bathroom is part of a larger house remodel, you'll want to involve an architect. Architects are the most highly trained of the pros you might hire. Look for one who specializes in residential design. A good architect will know how to meld your needs and desires into a plan that fits your budget and the architectural style of your home.

above • Who says tile has to end with a straight line? By literally thinking outside the box, somebody came up with this eye-catching undulating tile joint.

right • A soaking tub tucked into the corner of this sophisticated urban bathroom gives you the choice of watching TV or gazing at the cityscape beyond.

Personalized stools let kids reach the sinks while adding a red accent.

Should You Be Your Own General Contractor?

In almost all cases, the answer is no, even if you have advanced do-it-yourself skills. Smoothly pulling off a bathroom remodel requires careful coordination of subs. The carpenter, plumber, electrician, drywall installer, tile setter, and painter all need to appear at different stages of the process. If things get off schedule, a subcontractor is likely to tell you, "Sorry, but you had me scheduled for last week, so I can come back next month." If you have a full-time job outside your home, keeping things on track is going to be a challenge.

Beyond that, contractors have subcontractors that they work with regularly and have come to rely on. Those subs know that if they don't show up when scheduled, they may not get more work through the contractor. Your job might not be a top priority for a sub who knows you are unlikely to need him again for years, if ever.

Of course, there are exceptions. If you have the skills to pull off most of the work—say, you are comfortable with carpentry, painting, and basic wiring and you know a plumber you trust—of course you can save a bundle. You might want to check first with your family to be sure they don't mind the bathroom being out of commission while you take a month or two of weekends to do a job a contractor might complete in a few weeks.

In some cases, an architect may simply provide construction drawings for your contractor to follow or he may oversee the work. An architect may charge by the hour, or he may charge a percentage of the overall construction budget.

Bath designer. If you aren't planning major structural changes, you might hire a designer instead of an architect to help you envision your new bathroom. On many projects, bath designers work closely with architects as well as contractors and clients to make sure the use of space, color, and product choices fit the client's vision. In fact, many large residential architecture firms have designers on staff.

Look for a professional who has been certified as a bath designer by the National Kitchen and Bath Association (NKBA). There are several levels of NKBA certification—the highest is certified master kitchen and bath designer. A certified bath designer will have the training to create a floor plan, helping you make decisions such as where the fixtures will go and how big the shower should be. He or she will be up to date on trends in bath products from faucets to tubs to tile.

Interior designer. These professionals are qualified to perform the same functions as a kitchen and bath designer, but they don't specialize in kitchens and baths. As a result, an interior designer might not be as intimately familiar with the latest bath products as someone who focuses on kitchens and baths on a daily basis. On the other hand, if your bathroom is part of a larger home remodel, you might use an interior designer for the bathroom as well for the sake of continuity of vision—and perhaps to avoid too many cooks. In some states, interior designers must pass a licensing exam to use that title.

Many interior designers belong to the American Society of Interior Designers (ASID), and the ASID website can help you find a designer. Interior decorators can help you select furnishings and make other cosmetic choices. They don't have the formal training that interior designers do.

General contractor. A general contractor hires subcontractors for specific parts of the job while retaining overall responsibility for the work and scheduling. Some contractors do some of the work themselves—often the carpentry. If you find a contractor with lots of experience doing bathrooms, he or she may offer good suggestions for design, materials, and fixtures.

With a square tub as its centerpiece, this bathroom has a rectilinear motif that's softened a bit by a border of random stone.

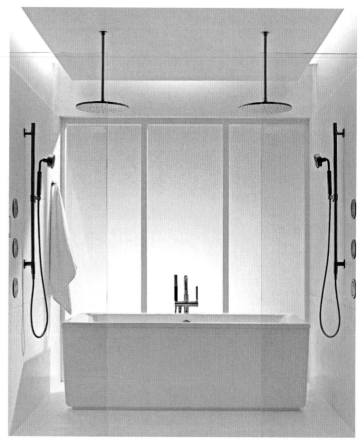

Here, a shower and tub are contained in a "wet room" with glass panels to prevent water from splashing into the rest of the bathroom. The shower is outfitted with rain showerheads, adjustable-height/handheld showerheads, and body-spray jets.

Artfully set in a corner with a beautiful view, this soaking tub doubles as a sculpture.

Some states require contractors to pass an exam and meet other criteria to be licensed. Other states require licensed contractors only for jobs costing over a certain dollar amount. And then there are states that simply require contractors to be registered, which is no indication of competency. To avoid work stoppage or even fines and removal of work done, it's important to know the requirement in your state. Homeadvisor.com lists licensing and registration requirements for each state, or you can check with your local building department.

When you do find a contractor you'd like to work with, ask to see proof of liability insurance, get his or her license or registration number if that is required in your state, and check that the number is valid.

Design/build firms. These are companies that combine design and construction under one contract. The idea is that if the builder and designer are already members of the same team, they will naturally work better and more efficiently together, saving the client money and making one entity responsible for all aspects of the job. Some firms are headed by contractors who employ designers or do design themselves. Other firms are headed by architects or designers who get involved in the construction work.

Small but Airy Casita Bath

When floor space is limited, look up. That's the secret to the airy feel of this tiny bathroom in a casita in Santa Fe, New Mexico.

The casita is attached to the main house by a covered walkway and serves as a guesthouse. It contains a shower, sink, and toilet in just 35 sq. ft. The room has just one window, yet it is bathed in light thanks to a large skylight over the shower. Narrow blue tiles with white grout cover the shower walls. The horizontal grout lines are random, while the vertical grout lines are continuous. The lines evoke a summer rain shower and draw the eye up the tall walls to the skylight, de-emphasizing the small footprint.

The tile continues around three walls of the room as a low wainscot, extending only about one-third of the way up the white walls, again emphasizing the high ceilings and distracting the eye from the small footprint. Omitting tile from one wall prevents the bathroom from feeling boxed in.

Mounted high on a shower wall is a waterproof industrial light fixture. "I use these fixtures a lot because they have a nice glass globe that seals them up so you don't have water issues, and they create a really nice shadow pattern," said Anni Tilt of Arkin Tilt Architects, the architect who designed the bathroom. This particular fixture has a cover that blocks light from shining up through the skylight. It was chosen because Santa Fe residents don't want light pollution to obscure their star-filled night sky.

The unusual sink was selected by the homeowner—Tilt originally had another sink in mind. "We were looking for something very small with a small countertop so you could put your kit bag on it. The client found this one and I was like, 'I don't know…' But it went so well in that space and we just went for it."

The casita is one open space, but to make it more versatile, there are two sliding panels that can be closed to partition off a bedroom area from the family room. As part of this plan, the little bathroom has two pocket-door entrances. One of these is blocked when the panels are in the open position. And when the panels are closed, the bathroom can be accessed from either the bedroom or the family room.

The unusual round corner sink provides a bit of counter space without crowding the door to the right or the shower at left. With no handle, the medicine cabinet looks like a framed mirror hung on the wall until you press on it to pop it open.

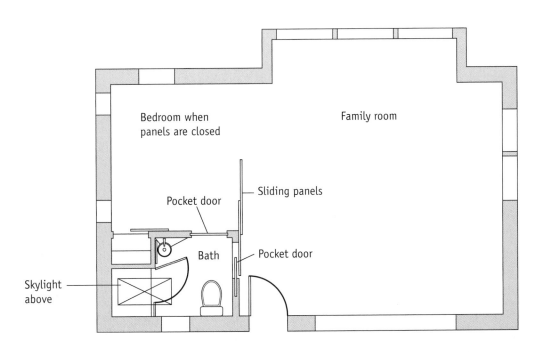

Bedroom when
panels are closed

Family room

Sliding panels

Pocket door

Bath

Pocket door

Skylight
above

above · Busy-looking floors don't work in small bathrooms. Here, large beige tiles with
white grout don't call attention to themselves and provide a nice visual contrast with the
much smaller shower tiles.

right · The skylight over the shower provides lots of natural light and gives the bathroom
an open feel.

Bathrooms that Work … for Everyone

You've probably heard the terms "accessibility," "universal design," "aging in place," and maybe even "living in place." But what do they mean? What's the difference between them? And why should you care?

An accessible bathroom is one that is designed to be usable by someone with a serious mobility issue, such as a wheelchair user. The Americans with Disabilities Act requires accessible bathrooms in public places but makes no requirements for private homes.

Universal design means creating spaces that are safe, comfortable, and accessible to people of all ages, sizes, and physical conditions from a 3-year-old to a 95-year-old to a 19-year-old who broke her leg while snowboarding as well as the three-quarters of Americans who wear eyeglasses or contacts. (You know, the silent majority who don't have a clue what is written on the back of a shampoo bottle!)

"Aging in place" expresses one of the goals of universal design—if you create a space that addresses the needs of older people, you'll be able to stay in your home as you age. The term "living in place" was more recently coined in response to market research showing that most people don't know what universal design is and don't want to think in terms of aging in place. Some contractors, designers, and other pros take courses with the National Association of Homebuilders (NAHB) to become Certified Aging in Place Specialists (CAPS) or with the relatively new Living in Place™ Institute to become Certified Living in Place Professionals (CLIPP™).

Whatever you choose to call it, it's simply smart to think about universal design when remodeling or building a new bathroom. Most likely your project will have physical and financial constraints that will require compromises. For example, you might not have room for a wheelchair-accessible shower, but if you are building a shower that happens to be large enough, consider making it curbless. You'll thank yourself if you ever

left and above • This bathroom was designed specifically to be accessible for a wheelchair user. It has a clear space under the sink, a low storage shelf, and a wide, curbless entry to the shower. Inside the shower is a built-in bench with a handy recessed shelf. An adjustable-height/handheld showerhead with a control is easy to reach from the bench or a chair. The handsome grab bars match the fixture finish. The stripe of smaller tiles with a blue border provides horizontal orientation as well as being an attractive design feature.

left • The shower in this gorgeous bathroom incorporates some important features of universal design. There's no curb, and the glass door swings inward away from the area that contains a bench and a handheld showerhead. The shower door handles are vertical so no one will be tempted to use them as a grab bar. The wide horizontal band of tiles features birds in flight.

need a wheelchair—even temporarily—and you'll eliminate a tripping hazard for everyone. Plus, curbless showers look great.

And looking great is a hallmark of successful universal design. Nobody wants an institutional-looking bathroom in his or her home. Fortunately, manufacturers realize this, so you'll have no problem finding gorgeous products that can be incorporated into universal design. Also, universal design isn't just about the products themselves—it's about smart use of products. Here are some ideas to get the conversation going when you talk to your architect or designer. Some of these ideas are simple enough to retrofit, so you might think about incorporating them in the other bathroom down the hall.

Think "thighs to eyes." That's about 24 in. to 60 in. above the floor—the range that's easy for most people to reach. When designing storage, try to keep the things you'll use most often at that height range.

Add a horizontal line. As mentioned, many of us wear glasses or contacts that we remove while taking a bath or shower. Add to that the fact that a shower fills with steam and we can lose a significant amount of depth perception if the wall is painted or tiled a consistent color. Creating a band of contrasting color—paint or a course or two of tiles at about 5 ft. from the floor—provides depth perception and a level orientation. It can look really sharp, too; in fact many bath designers use horizontal lines to tie elements together visually or to make a bathroom seem more spacious. If you have glass shower doors, a horizontal line etched into the glass is also helpful.

Make shower controls easy to reach. As larger showers have become more common, bath designers have begun putting the controls near the entrance rather than directly under the showerhead so you don't have to lean over to avoid the water while adjusting the temperature. Turns out, this is also better for your back and decreases the chance of falling as we get less nimble with age.

Install a curbless shower. Showers that have no curb and employ a linear drain have exploded in popularity simply because they look better and more up to date than a shower with a curb and round drain in the middle. A shower without a curb is safer for everybody and can make the shower accessible to a wheelchair user. Creating a custom curbless shower with a linear drain can be expensive because the slope needs to be built into the floor. But manufacturers have responded with sloped shower pans that bring the cost down. Keep in mind that a curbless shower needs to be at least 36 in. wide to prevent water from splashing out.

Use vertical shower door handles. A horizontal shower door handle is dangerous because people may grab for it if they slip. Shower doors are not designed to take weight and could come down if grabbed.

Make sink controls easier to reach. This is a good example of the difference between accessibility and practical universal design: If you were designing for true accessibility for someone in a wheelchair, you would leave a clear space under the sink so the person could roll his legs underneath. However, if you are not designing specifically for a wheelchair user and you don't want to give up the storage a vanity supplies, you might position the vanity in a spot where someone could parallel park their wheelchair next to it and you could place the controls at 2 o'clock or 10 o'clock so the person could reach them. You can accomplish this by rotating a round sink with faucet holes or you could use a sink with no holes in conjunction with controls mounted in the counter. The sink will be easier for kids to use, too.

Install single-lever sink faucets. These decrease the probability that a user will accidently turn on full-hot water. Levers in general are easier for children and people with arthritis to use.

Use door levers. As with sinks, levers are easier to operate than knobs. Plus, you can push a lever down with your elbow when you enter the bathroom with an armload of towels. It's also helpful for door handles to be in a color that contrasts with the door.

Carefully consider toilet height. Most experts recommend installing "comfort-height" toilets that are 18 in. high because they make it easier for older people to sit down and get up. But Richard Duncan, executive director of the Universal Design Institute, points out that comfort-height toilets can be difficult for short people, including children, to use. Noting that there are devices to raise toilet seat height, Duncan suggests that a comfort-

above and right • **This sink is lower than usual height with space underneath to make it accessible to a wheelchair user. Doors slide back under the counter so they won't get in the way of a wheelchair.**

above and right • **Increasingly, manufacturers are producing grab bars that match faucets and other fixtures and handholds that are incorporated into other functions, such as a toilet paper holder or the towel rack shown here.**

height toilet can be a good idea but only if you have a lower toilet elsewhere in the home. Wall-mounted toilets can be placed at any height you like.

Install an outlet behind the toilet. Today you can buy toilet seats that are heated, provide washing and drying functions, and incorporate a nightlight. These seats need to be plugged in. You might not opt for a multi-function seat now, but all of those functions are helpful for older people and people with mobility issues. Adding an electrical outlet is cheap and easy during construction, and it is also a great place to plug in a simple nightlight.

Provide a strategic nightlight. A nightlight behind the toilet or incorporated into the toilet seat is ideal because it pinpoints the most likely middle-of-the-night destination. An LED strip installed in a vanity kick is another easy-to-install solution.

Provide grab supports. The word "grab bar" conjures images of institutional stainless steel that nobody wants in their bathroom. But you can find grab bars in a wide variety of colors, styles, and materials to match any décor. An attractive grab bar wrapping around the bathroom can be a striking design feature. You'll also find towel bars, toilet paper holders, and cup holders that have short, small-diameter handholds inconspicuously built in. There are even sinks with handholds molded into the sides. It is a very good idea to provide a short handhold directly over the shower controls. People will get used to grasping the handhold with one hand and adjusting the controls with the other. And should someone slip, they'd reach for the handhold, not the controls.

Put switches at 44 in. In many homes, the bottom or the top of light switches are positioned 48 in. from the floor. This was done simply because drywall sheets are 48 in. wide and drywall installers found it easier to notch either the top of the bottom sheet or the bottom of the top sheet than to cut holes. However, centering the switch at 44 in. puts less strain on the shoulder. Most drywall installers now cut switch holes after putting up the sheets, so placing switches at the optimum height is no extra work.

Heat the floors. Many people install radiant floor heating in the bathroom simply because it is a pleasure for bare feet. It's also safer because heated floors dry faster, and dry floors are less slippery.

Conserving Water and Energy

The bathroom is the biggest consumer of water in the house, with the toilet alone using up to 27% of the 400 gallons a typical family of four uses each day. It takes energy to heat water, and it's easy to forget that we use electricity to move that water around. If you use municipal water you are paying indirectly; if you have your own well you are paying to run your water pump. It all adds up to a lot of water—and dollars—down the drain.

There are two overall strategies for using less water—you can use water-efficient fixtures and you can install an efficient plumbing system. The first is very easy to do, especially if you are building or remodeling. Just make sure to install showerheads, faucets, and toilets that have the WaterSense® label. WaterSense is a voluntary partnership program sponsored by the U.S. Environmental Protection Agency. Partners include manufacturers, retailers, and distributors of products that use water. To qualify for the label, a product must be at least 20% more efficient than average products in that category. Plus, qualifying products must perform as well or better than other models.

The best plumbing design is an integrated system built into the house and involving the kitchen and laundry as well as the baths. For example, you could opt for "home run" hot water supply lines in which each fixture or faucet has its own pipe directly to the water heater. This will cost more to install than a more conventional trunk-and-branch layout, but you'll lose less heat as the water travels through the pipes, saving money in the long run. You'll also save time and water because you won't have to run water as long while you wait for hot water to reach faucets.

top right • A fold-up seat and adjustable-height showerhead are great examples of universal design.

right • This tub has a pull-up side, so a wheelchair user can slide over into the tub.

But it's probably not practical to retrofit an entire water distribution system unless you are building a new house or gutting an old one. One good option for a bathroom remodel or addition is to install an auxiliary hot water heater in or near the new bathroom—there are small models designed to fit in a vanity. This would eliminate the wait for hot water. You can also install an on-demand water heater that doesn't provide instant hot water but does provide limitless hot water.

One very simple energy-saving strategy is to prevent hot water from cooling by insulating water pipes. If your water heater is warm to the touch, purchase an insulated cover made to fit it. More complex strategies such as a drain-water heat recovery system cost more initially but can also reduce hot-water costs substantially.

Reusing gray water that goes down the drain after a shower or from a washing machine also holds promise for using less water. While gray water isn't potable, it's not especially dirty either and can be used for watering the lawn or garden. If code allows, gray water can also be used for flushing toilets.

All of these things are worth discussing with your plumber when you are building new or undertaking a major bathroom remodel. You'll find that taking relatively simple steps while the house is under construction may pay dividends down the road.

top left and left · **This bathroom is designed for all users. The low sink at left with doors that fold in out of the way and a wide curbless, doorless shower entry provide easy wheelchair access. The sink at right is at the typical height over an enclosed vanity.**

Mirror Images Take Advantage of Expansive Views

This house in the Taconic Mountains of New York has magnificent views of rolling hills in three directions—the Catskill Mountains to the south, the Adirondack Mountains to the north, and the Berkshire Hills of Massachusetts to the east. This vantage point was so important to the homeowners that to achieve it they moved an old Cape Cod–style house from another part of the property and then built a large addition.

"The whole idea of the house was to find ways of enjoying the landscape, so the house is really more about the property than it was about being inside a building," said architect Michael Waters of LDa Architecture & Interiors.

The homeowners wanted the utility of two separate bathrooms but they didn't want to divide the space into separate rooms that would interrupt the views. And so, Waters worked with the design firm Weena and Spook to develop a design that, in two senses, uses the idea of mirror images.

In one sense, the east and west walls of the room are mirror images. Each side has an identical water closet and walk-in shower with an identical window between. In the middle of the room are back-to-back vanities separated by a mirror. In another sense, the design is literally about the images in the mirror—because the two sides of the room are identical, the images in the mirrors make the mirrors seem to disappear.

This strategy left the north wall of the room open for two double-hung windows that are so enormous that you can easily walk through them when open. And because the windows face north, the light is never overpowering as it passes east to west over the house. Clear glass in the shower stalls that flank the wall doesn't interrupt the light and creates dynamic reflections.

"We were very interested in the reflectivity of the mirrors obviously, but also the counters, which are polished, and the glass showers, which are both reflective and transparent," Waters said.

The wooden frame dividing the two vanities contains a mirror on each side. Because the frame exactly divides identical sides of the room, the frame appears empty, contributing to the openness of the space.

right • Each identical side of the bathroom has a glass-enclosed shower that seems to disappear into the corner. Each side also has a wood-panel cubicle that houses a toilet and small wash sink. The cubicles are exactly the same size as the shower enclosures so that the two elements complement each other even as they offer contrast.

bottom right • A dressing table occupies the wall opposite the huge north-facing windows. Small light fixtures spaced around the sides and top of the mirror frame provide excellent shadow-free light.

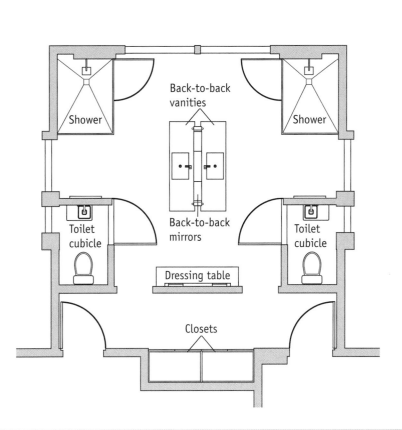

SHOWERS

Showers are becoming larger and taking on custom shapes,

while innovations in showerheads help enhance the spa-like experience.

AND

Separate tubs also come in more configurations than ever before,

including deep soaking tubs that make bathing more luxurious.

TUBS

Showerheads and tubs that play music. Deep soaker tubs with lights that change color to suit your mood. Showers with multiple showerheads and steam. Gone are the days when almost all bathrooms were equipped with a 5-ft. tub with a showerhead and curtain. Nowadays, you have lots of options for showers and tubs as the bathroom continues to morph from a place we simply go to get clean into a retreat where we linger and relax. And while spacious bathrooms certainly are nice, you can enjoy many of the latest amenities even if your bathroom footprint and budget are limited.

If your space is indeed limited, the first question you might ask yourself is whether you need both a shower and a tub in the same room. You'll most likely want a tub somewhere in the house, especially if you have kids and—even if you don't—to protect the home's resale value. But many folks are opting to forgo a tub in favor of a more luxurious dedicated shower space, especially in a master bathroom. A shower is, after all, the fastest and most efficient way to get clean and the top choice for most people, especially on workdays.

A soak in a whirlpool or air-jet bath can be a wonderfully relaxing experience. However, big tubs use a lot of water and take time to fill. Many people just don't find the time to use them much. If you don't take many baths, consider investing your money and space in a larger, more luxurious shower.

For those who have space, a separate tub and shower in the same bathroom has become the most popular choice. Often the shower enclosure is all glass, with the room's floor and wall tile patterns running through the shower space. This makes the bathroom seem more open and spacious. Curbless shower floors enhance this effect and also make the shower space more accessible.

Another popular option is to use glass to partition the shower and the tub into a "wet room" area of the bathroom. With this strategy you can tile the wet room and use other less waterproof materials, like wooden wainscoting and floors, in the rest of the bathroom.

Where space allows, many people now opt to separate the bath from the shower—it makes both bathing and showering more pleasant. Here, the tub surround flows past the glass to become a seat in the shower.

In this modern minimalist bathroom, the tub and shower are in their own "wet room," separated from the rest of the bathroom by a partition and door of frosted glass. This protects the wooden details from excess moisture while making the wet area easy to clean.

Because the floor tile and wall colors are continuous and the enclosure is glass, this shower doesn't visually interrupt the space, making the bathroom seem larger. The shower is curbless and it doesn't have a door—two factors that improve accessibility.

Showers

Showers offer a lot of design flexibility in size, space, and aesthetics. The huge variety of stone, ceramic, and glass tile and glass block along with the availability of expansive panes of tempered glass means you can create a shower space that makes its own statement or beautifully blends into the overall bathroom design.

Some showers in master baths are quite large, sometimes big enough to accommodate two people. But a shower need not be huge to be luxurious—you can still opt for a showerhead that simulates rain or one that pulses or plays music, and you can still install steam.

above · This shower features a unique sitting nook with a window. The linear drain in the left side of the floor has become a popular alternative to a traditional round center drain underfoot.

left · A very private setting offered the opportunity to blend this shower into the outdoors. The color and texture of the wall and floor tiles are inspired by the evergreens beyond. The tiles stop short of the ceiling, suggesting the sky over the trees. In keeping with the theme, a large rock serves as a seat and there is a rain showerhead. Clear glass doors let the homeowners enjoy the light and view from anywhere in the bathroom.

facing page left · A lot of punch is packed into this relatively small shower, including fixed and handheld showerheads and steam. Controls are located at the entrance to the shower so the user doesn't need to dodge the shower spray while adjusting temperature. Two rows of glass block provide light without sacrificing privacy.

COMBINATION TUB/SHOWERS

While the tried-and-true bath/shower combination has been joined by other options, it's likely to remain a mainstay in many American homes. It can be purchased as a prefabricated plastic tub/shower unit or created on-site with a standard tub and tile surround. In period homes, a simple curtain rod encircling a claw-foot tub may be all that is needed.

Whatever form it takes, a bath/shower combo is a cost-effective and space-saving way to combine two functions into one fixture. For example, it's a smart choice for a small bathroom that will be used to bathe kids but also needs a shower for guests. Besides, even if you don't often take baths, everyone could use a therapeutic soak once in a while.

Prefab combo units have the advantage of being quicker to install—you'll spend less on installation and materials than you would with a standard tub and tile surround. With no grout or caulk joints, prefab units require less maintenance and are easier to clean. However, prefab units don't offer the vast choice of colors and texture that tile provides.

top right • A simple curtain on an encircling rod, exposed plumbing, and a traditionally styled freestanding tub add up to old-timey elegance.

right • Here is an elegant way to make a small space seem larger. Rather than the typical three-walled tub/shower surround, the wall opposite the showerhead is left open except for a short ledge. This is practical because the rain showerhead directs water straight down. A short glass door enhances the open look and doesn't swing far into the cramped space. The glass enclosure admits natural light from the window.

facing page • Prefabricated tub/shower enclosures are economical, practical, and durable, and some models have lots of shelves molded in. With no grout or caulk joints to worry about, these units are also easy to clean. They are typically less expensive, leaving you money for other details like these built-in shelves.

CURBLESS SHOWERS

With a tub/shower combination your design choices are limited by standard tub sizes. However, a stand-alone shower can be any practical size or shape that fits your design. Eliminating the curb gives you even more design flexibility—curbs get complicated if they are not straight and short.

Perhaps the main reason curbless showers have become popular is simply because they are a handsome design feature—a continuous tile floor makes the room seem larger and gives it a pleasing flow. But there are some very practical advantages as well. For one thing, it's easier to mop a floor that's not interrupted by a curb.

Most importantly, a properly designed curbless shower is safer (nothing to trip over) and more accessible. An uninterrupted floor combined with a doorless entry, grab bars, and a lower handheld showerhead can make a shower accessible to wheelchair users. And even if accessibility isn't a concern now, a curbless shower can be more easily adapted to future needs. See pp. 20–23 for more on accessible design.

above • The countertop continues into the shower while the shower wall and floor tiles continue into the room. The result is a pleasing flow of elements.

facing page • In addition to being practical, a curbless shower can provide a design opportunity. The small floor tiles in the shower area add texture and also mimic the color of the tiles behind the tub.

Framing Requirements for Curbless Showers

Conventional showers have a shower pan that slopes toward a drain in the shower itself so the shower can be installed over standard level floor framing. Installation gets more complicated with a curbless design because there is no shower pan. The floor itself must slope toward a drain that serves the shower or whatever area of the floor you decide will be part of a "wet room." Otherwise, you risk having shower water creeping under the bathroom door into the room or hall beyond.

Unfortunately, incorporating the required slope into the floor framing is not always feasible, especially in renovation projects. If you are considering a curbless shower, be sure to talk with your architect and/or builder during the design phase of your project. The time to broach the subject is not when the floor has been framed and the plumber or tile setter is about to start work.

Niches

Recessed niches are most typically found in showers, but they can be handy next to a dedicated bathtub, too. They typically have one or two shelves to provide convenient storage for shampoos and soaps. Beyond that, they break up fields of tile, providing an opportunity to add visual interest. A niche can incorporate tiles that contrast in color, texture, and/or size with the field tiles, or for a subtler look, they can match the tiles.

The simplest niches are preformed boxes designed to fit between two studs and then covered with tile. There are also preformed boxes designed to span more than two tiles—these require some extra framing to take the load of the interrupted studs. And, of course, boxes can be custom-made to create niches of any dimensions, even spanning an entire shower wall.

However it is done, the installation must be waterproof. A leaky niche will allow water to get behind the tile backerboard, inevitably causing rot. The tile setter must take great care to choose the right materials and install them properly.

Unless the wall has tiles of different sizes with random grout lines, the niche should be laid out so its edges line up with grout lines in the rest of the tile wall. This looks more finished than having the sides of the niche land haphazardly on the field tile. If that's not possible, a border can make the problem much less noticeable.

1. Because the thin, multicolored tiles used in this shower are so eye-catching, the designer chose to use the same tiles in the niches rather than interrupt the effect. 2. A large niche with a shelf provides plenty of storage for everybody's shampoos, conditioners, and soaps. 3. Tub-side niches don't get soaked, offering the opportunity to create a warm wooden detail. This one saves space in the room by incorporating the tub spout and controls and even a small cabinet.
4. The striking contrast of black-and-white tiles is enhanced by a niche that includes a light fixture along its length.
5. The tile in this niche matches the contrasting tile used in an adjacent wall. Because of its size and dramatic color difference with the white subway tile, the niche becomes a focal point.

Showerheads

As luxurious showers have overtaken whirlpool baths in popularity, manufacturers have responded with an increasing variety of ways to spray water at you. In addition to the familiar showerhead mounted to a pipe coming out of the wall, there are showerheads that slide up and down a pole to accommodate people of different heights. Sliding showerheads often are detachable so they can be used handheld. And however they are mounted, many have sprays that are adjustable from mist to a more powerful blast, sometimes pulsating. Or if you prefer, you can install a showerhead that mimics the effect of rain. This can be mounted to the ceiling or on a longer pipe high on the wall.

Can't decide? You can always install multiple showerheads, but if you intend to use them at the same time, you'll need to make sure your water pressure and water heater are up to the task. An EPA requirement limits showerhead water volume to 2.5 gallons per minute. If a device bears a WaterSense label, it uses 1.5 to 2 gallons per minute.

A more practical reason to install a second showerhead is if the bathroom will be used by someone in a wheelchair or who needs to be seated while showering. A second, lower showerhead is nice for kids, too. As long as the heads aren't being used at the same time, you don't need to worry about extra water consumption.

To make things more interesting, you can add body sprays, which are strategically mounted at several places along the wall to shoot massaging jets of water. Many fixtures allow you to aim the jets. Here again, you'll need to consult with your plumber to make sure there's enough water pressure to make the jets effective and enough hot water capacity to last through a shower.

Recognizing that installing multiple shower devices calls for complicated plumbing, several manufacturers have developed shower panels that are simpler to install. These panels have various combinations of showerheads and jets mounted on them. You can buy a panel that includes a rain head, a wall-mounted head, and jets and a spigot for filling a tub or bucket all controlled by a selector knob. A panel can be less expensive than buying several individual fixtures, and it will save you money on installation.

above · **This pendant from Brizo® offers a different take on overhead rain showerheads.**

facing page top left · **This multifunctional showerhead is made even more versatile by an optional jointed arm.**

facing page top right · **A handheld showerhead mounted on an adjustable holder is a very versatile option, especially worth considering if shower space is limited.**

facing page bottom left · **Users have lots of options in this shower built for two: wall-mounted showerheads, ceiling-mounted rain showerheads, a handheld sprayer, and a total of six wall-mounted body spray units.**

facing page bottom right · **Several manufacturers make showerheads that use Bluetooth technology to play music.**

WHIRLPOOL TUBS

Jets of warm water act like a massage, easing the ache of sore and tired muscles and relieving the stress of everyday life. The appeal of a whirlpool bath is easy to see, and many homeowners wouldn't be without them. Yet members of the NKBA reported in a recent survey that 58% of their remodeling projects involved *removing* a whirlpool or tub. Oftentimes, homeowners replaced a whirlpool with a non-whirlpool tub or a shower.

Because of its pump and piping, a whirlpool tub takes up more space and is more expensive to purchase and install than a conventional tub. And most whirlpool models take more maintenance than other tubs because water remains in the pipes when the jets are off. This allows bacteria, dirt, soap, shampoo, and body oils to build up. The maintenance is pretty simple—you cycle through some dishwasher detergent every month or so. Still, some folks have found they just didn't use the whirlpool enough to justify the extra expense, space, and maintenance. Another consideration: Does your water heater have enough capacity to fill it?

So give it careful thought. If you decide you want a whirlpool, you'll find many brands and types on the market. In some tubs, the jets can be adjusted for both direction and water volume. The number and location of jets varies by manufacturer. Tubs with in-line heaters will keep the water warm while you are relaxing.

AIR-JET TUBS

Air-jet tubs shoot jets of preheated air into the tub, creating a massage-like experience that's gentler than the water jets of a whirlpool. The jets of a whirlpool provide more intense pressure than the tiny air-jet bubbles. A whirlpool is better for targeting the neck, back, and other parts of the body that need stress relief. Air jets are said to be better for stimulating blood flow and breathing.

Some tubs let you adjust the intensity of the air jets, and some have a wave-and-pulse feature that works by presetting the jet intensities to change according to a pattern. Some models have in-line heaters.

Because there is no residual water in the pipes, air-jet tubs don't require as much maintenance as most whirlpools. Plus, air-jet tubs are not as complex as whirlpools, so they are less prone to problems. If you are deciding between whirlpool and air jet, it's a good idea to visit a showroom that has working models of both.

JAPANESE-STYLE SOAKING TUBS

Any tub deep enough to immerse yourself might be considered a soaking tub. However, the traditional Japanese-style tub, called *ofuro*, which simply means bath, is distinctive because it's designed to immerse you up to the neck while sitting up. As a result, these tubs are quite deep but have a smaller footprint than western-style tubs.

Because of the smaller footprint, an *ofuro* can be just the ticket if you want to fit a soaker into a tight space. However, due to the depth, you'll typically need to provide a step up into

Air-jet tubs provide a relatively gentle full-body massage, while water jets provide a more intense targeted massage.

LED lights change the water color in tubs equipped with chromatherapy.

Some Japanese-style soaking tubs are designed to be installed into a platform. Others, like this one, are freestanding and meant to be celebrated as a sculptural focal point in the bathroom.

the tub or recess the tub partway into the floor. Like whirlpools and other large tubs, *ofuro* tubs hold a lot of water so you'll need to make sure your water heater has enough capacity. Also, because the weight of the water is concentrated on a relatively small floor area, you may need to reinforce the floor joists.

Ofuro tubs have become popular around the world and are widely available in the United States. They are made from a variety of materials including wood—the traditional material in Japan—acrylic, stainless steel, and copper. You can have one custom-made from concrete.

COLOR- AND SOUND-THERAPY TUBS

Some whirlpool or air-jet tubs are available with a color therapy, or "chromatherapy," lighting feature. Chromatherapy is a form of holistic medicine that uses color to promote physical and mental well-being. Tubs with this feature incorporate light-emitting diodes that, with the push of a button, produce a range of colors. Typically the lights will automatically cycle through the colors, but you can set them to one color that suits your mood.

If you want to involve yet another one of the senses, you can get a tub that offers vibro-acoustic therapy—the idea is that feeling musical vibrations is therapeutic. Kohler, for example, sells this feature as VibrAcoustic®. You use Bluetooth to send your music from your phone or other device to the tub. When the tub is empty, it simply plays music, when you are in the filled tub, you feel the vibrations through the water as well as hear the music.

TUBS WITH DOORS

Older people and those with a disability may find it difficult to get in and out of a conventional bathtub. One solution is a walk-in tub with a door. With the door open, the bather steps over a very low curb, takes a seat, closes the door, and fills the tub. Models are available with whirlpool jets, air jets, or both, and some come with handheld showers.

Tubs are available with doors that swing in or doors that swing out. An in-swinging door would make sense in a cramped bathroom where clearance is a problem. One argument against an in-swinging door is that they are harder to exit in an emergency because most of the water has to drain out before you can open it. However, in its favor, an in-swinging door is pushed against its seal by the weight of the water, making a leak unlikely.

Like other deep soakers, walk-in tubs are typically more compact than conventional tubs.

Walk-in tubs with doors are helpful for people who have difficulty getting in and out of conventional tubs.

Bathtubs

When it comes to durability and appearance, the material used to make your bathtub is more important than the shape or features like whirlpool jets. You have lots of materials at different price points to choose from, ranging from fiberglass or porcelain over steel at the low end to copper, bronze, and stainless steel, which are extremely durable and quite expensive.

FIBERGLASS/GELCOAT
$

- A mold is sprayed with a thin layer of polyester resin called gelcoat, and then layers of fiberglass are added for strength. However, the tub walls are thinner than other materials so may not seem as solid.
- Resists stains, but the topcoat is thinner and not as durable as other plastic options, such as acrylic.
- Color may fade over time.
- Fiberglass surface is easily scratched, so abrasive cleaners should not be used. Follow the manufacturer's cleaning instructions.

ACRYLIC
$$

- Made by forming sheets of acrylic in a vacuum mold and reinforcing the tub with fiberglass.
- More durable than fiberglass/gelcoat, with a thicker color layer so minor scratches won't be obvious.
- As with fiberglass/gelcoat, abrasive cleaners should not be used on acrylic.

PORCELAIN OVER STEEL
$$

- Porcelain enamel is fused with heat to a substrate of sheet steel formed into a tub.
- Lighter in weight than cast-iron tubs but not as durable.
- Loses heat quickly.
- Porcelain is easy to keep clean, but any nicks or cracks in the surface will allow the steel tub to rust.

CAST POLYMER
$$-$$$

- Made by casting plastic resins with filler. The filler may be natural stone, quartz, or marble to mimic the look of solid stone. Or the filler may be a synthetic material to produce solid-surface material in a variety of colors. Usually the surface is a layer of gelcoat.
- Gelcoat surfaces can be damaged by abrasive cleaners.
- Some manufacturers will make sinks and shower enclosures at the same time to ensure a good color match.

PORCELAIN OVER CAST IRON
$$$

- Porcelain enamel is fused with heat to a thick substrate of cast iron.
- The porcelain layer is thicker and more durable than a sheet-steel tub.
- Cast iron is heavy, adding to installation cost. The floor may need reinforcement.
- Because of greater mass, cast-iron tubs retain heat very well. However, that greater mass will cool water if the tub is cold when filled.
- More durable than fiberglass/gelcoat, with a thicker color layer so minor scratches won't be obvious.
- Like fiberglass/gelcoat, porcelain should not be cleaned with abrasive cleaners.

COPPER, BRONZE, AND STAINLESS
$$$-$$$$

- Very expensive but also extremely durable.
- Like cast iron, these metals retain heat well but will cool water if filled while cold.
- Not susceptible to corrosion, but some metal surfaces require polishing to maintain their original luster.

This serene bathroom exudes order and simple elegance.

Tub Fillers

Style-wise you'll most likely be choosing tub controls and spouts to coordinate with your sink faucets—they are available in all the finishes discussed in "Faucets," p. 80. Aesthetics aside, all tub fillers fit into one of three categories: deck-mounted, wall-mounted, and floor-mounted. And like faucets and sinks, the type of tub filler will depend on the type of tub.

DECK-MOUNTED

These tub fillers are designed to be mounted in one to five holes on a horizontal surface—either through holes in the rim of the tub itself or in a constructed deck. One-hole models have the controls integrated into the spout; two-hole models have a spout and one control for both hot and cold water. Three-hole fillers have a separate spout plus two controls. Some models use the second or third hole for a handheld shower, and there are also models that place the handheld shower in a fourth hole. Some even require a fifth hole for a control that diverts the water between the fixed spout and the handheld shower. Handheld shower spouts are available for wall-mounted and floor-mounted fillers as well.

WALL-MOUNTED

If your tub has no holes and is next to a wall, you can use a wall-mounted filler. These come in a variety of configurations and styles and can add a very elegant built-in look. Plus, installing one just might save the bit of space you need for a tight tub installation.

FLOOR-MOUNTED

Tub plumbing coming up through the floor evokes images of old-fashioned claw-foot bathtubs, and there are many traditional-looking floor-mounted tub fillers available. However, you'll also find plenty of streamlined and contemporary-looking models designed to complement contemporary-looking freestanding tubs. Floor-mounted fillers are generally more expensive than those mounted on the wall or floor simply because more finish is needed to cover the exposed pipes.

This sleek tub filler includes a handheld sprayer. It's designed to fit a contemporary-style tub with four holes in its narrow rim.

above • An old idea made new, this floor-mounted tub filler with sprayer wand mounted to the side is right at home in this very modern bathroom.

right • Simple wall-mounted chrome filler spout and handles complement a contemporary slipper tub.

facing page bottom • This classic-looking deck-mounted tub filler includes a handheld sprayer in a cradle that's reminiscent of an old-time telephone.

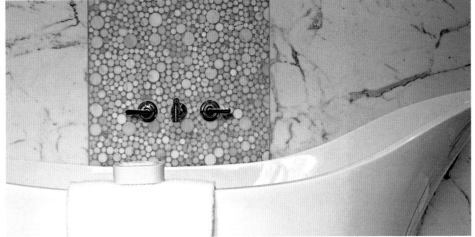

Angles Make a Small Bath Seem Spacious

How do you fit a double vanity, a walk-in shower, a separate tub, and a toilet into 110 square feet? Plus, you want lots of natural light despite being on the street. And, oh yes, you want plenty of privacy, but you want an open, spacious feel with no doors.

These were the challenges architect Sophie Piesse faced when she designed a house on a town lot in Carrboro, North Carolina. The house has three 14-ft.-wide sections forming a U around a courtyard. On the first floor, one leg of the U comprises the master bedroom, a large walk-in closet, and the master bathroom. When you walk into the master bedroom you can't see into the bathroom or the closet, both of which have no door. Walk through a short hallway and you'll find the bathroom on your right and the closet to your left. Inside the bathroom, the toilet is tucked unobtrusively into a nook on the right.

Several visual devices make the bathroom seem larger than it is. First, a ledge runs over the vanity and turns to become sills for the two windows that meet at the corner. This strong horizontal line leads the eye and ties the whole space together. The vanity floats, providing another strong horizontal line at its bottom without interrupting the floor. Even the drawers contribute to the horizontality. The grey tile and countertop seem to recede rather than emphasize the walls.

The window blinds close from the bottom up to provide privacy from the street without completely blocking the view. Even fully closed, these translucent blinds admit light.

The graceful and slim soaking tub seems an obvious choice for a small space, but the decision to place it at an angle is less intuitive. That decision was both practical and aesthetic, Piesse notes. "It gives you a little more access to the shower," she said. "Also, because the windows are asymmetrical, it worked to have the tub asymmetrical."

Because turns, not doors, are used to create privacy, the views keep changing as you move through the space. "Tweaking the tub added to that sense of movement," Piesse says. "It's all very much about line of sight and ease of access."

The toilet is tucked into a corner next to the vanity, which is one reason a bathroom door was deemed unnecessary.

The walk-in shower has a pane of glass to contain the shower spray, but no door. The shower space is tapered just enough to prevent access to the shower from seeming crowded by the tub.

Strong horizontal elements help make this small bathroom seem more spacious. These elements include a ledge that runs along three walls, the long vanity and countertop, and the continuous lines of the wide drawers. Even the blinds, which pull up instead of down to provide privacy when partly shut, add horizontality. The ledge works with the tub set at an angle to provide a sense of flow through the space.

A dresser built into a wall just outside the bath is in the same style as the vanity, contributing to the continuity of the space.

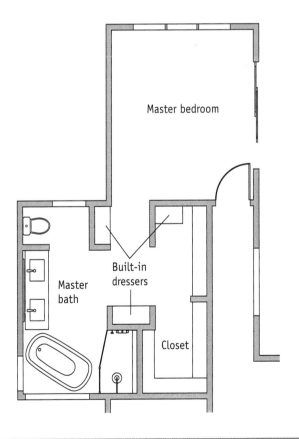

HEATING,

Choosing the right mechanical systems for your bathroom

is key to your health and comfort.

COOLING &

It's also essential to preventing moisture damage

inside the walls and roof.

VENTILATION

When it comes to climate control, bathrooms have needs beyond those of other rooms in the house. The major difference is, of course, the need to ventilate odors and humid air caused by showers and baths. As homes have become more airtight, a properly designed ventilation system is crucial to providing healthy air and preventing mold and other damage caused by condensation.

There are several approaches to ventilation—the one you choose will depend on your budget and whether you are building new, renovating, or remodeling. The simplest approach is an operable window, and, in many areas, building codes require no additional ventilation. As a result, many bathrooms, even in newer homes, have no active ventilation at all. However, the reality is that nobody wants to open a bathroom window when it's cold outside or when the space is air-conditioned. So even if you're doing a light remodel that doesn't involve opening walls, it's a good idea to include at least a simple fan in the ceiling or wall to exhaust air to the outside.

Exhaust fans are the least expensive form of active ventilation, and a properly sized fan can do an excellent job of removing humidity and odors. The drawback is that you paid good money to heat or cool the air you are now sucking out of the bathroom. For this reason, you may want to invest in a heat- or energy-recovery ventilator (see "Energy-Saving Ventilation," p. 149).

Whatever ventilation system you choose, it won't be effective unless it gets used. Many people simply forget to turn on the fan or want to avoid the noise, even though very quiet fans are now available. And even if a fan is used, it typically needs to remain on for about 20 minutes after you leave the bathroom. So rather than a simple on/off switch, you want to choose a switch that automates the process as much as possible (see "Switches and Timers," p. 148).

Even powder rooms without showers can benefit from an exhaust fan. Moisture may not be a big issue, but powder rooms are usually located near main living areas, so it is important to vent odors.

In most cases, heat, and often air conditioning, for the bathroom will be provided by the home's central system, whether you are building new, renovating, or remodeling. Still, because we are often unclothed in the bathroom, the space may benefit from some spot heating, such as an infrared overhead lamp just outside the shower or in-floor heating to warm bare feet.

Mechanical cooling isn't essential in all parts of the country, but in most areas, it does make life more bearable. If your home is heated by forced hot air, you may be able to retrofit air conditioning through the existing ducts. But if your home is heated by a hot-water system without ductwork, installing central air conditioning in an existing bathroom is likely to be impractical. However, a type of air-to-air heat pump called a "ductless mini-split" makes it possible to add cooling (and heating) to just a few rooms without conventional ducts. It's an option worth exploring in both new construction and remodels.

top left • Building codes typically don't require mechanical ventilation in a room with operable windows. However, even if the windows will adequately ventilate, you don't always want to open them. An exhaust fan is always a good idea in a bathroom.

left • Hot-water pipes or electric wires under the floor are an excellent way to heat the room while warming tiles that otherwise might chill bare feet. Hot-water radiant heat is best for large spaces and is usually installed as part of a whole-house system. Electric resistance heat is more suited to supplementary heat in smaller bathrooms.

143

Fans

There was a time when your typical ventilation fan sounded a bit like a jet taking off. No wonder folks were reluctant to turn them on. But nowadays you can purchase a bathroom exhaust fan that is all but inaudible—if anything, you might forget to turn it off! And, if you want to hide the exhaust fan from the eye as well as the ear, you'll find a variety of ceiling-mounted light fixtures with an exhaust fan incorporated into them.

If you don't mind seeing the fan's intake grill, you can save money by purchasing a unit that is just a fan or one that has a simple unobtrusive light source incorporated into it. There are also ceiling-mounted units that combine a fan with an infrared heater.

Because vapor-carrying hot air rises, the ceiling is the most effective place to install an exhaust fan. It's also less conspicuous than a wall-mounted fan. But if there is no practical path for venting through the ceiling to the outside, a fan mounted on the wall near the ceiling is nearly as good.

The most common and basic fans have the fan and motor in a housing that is mounted flush to the wall, though you can also get an in-line fan that is mounted elsewhere in the ductwork. These fans can draw air from more than one intake; however, they are more expensive and often noisier than ceiling- or wall-mounted fans.

Whatever style of fan you choose, it is important to purchase one that is capable of moving enough air to do the job as measured in cubic feet per minute (CFM). Code requires at least 50 CFM. The Home Ventilating Institute (HVI) recommends at least 1 CFM per square foot of floor area up to 100 square feet. For bathrooms larger than 100 square feet, the HVI recommends 50 CFM per fixture—so if your large bathroom has a shower, a toilet, and a separate tub, you'd want a fan rated for 150 CFM. Fans are most effective when they are located directly over the source of moisture, so a large bath with separate shower and tub might have separate fans for each. Also, if the toilet is enclosed in a space with a door, it should have its own fan.

In any bathroom exhaust-fan installation, it's crucial that the air be ducted directly outside. Never vent a fan to the attic, basement, or other interior space where moisture can condense on cold surfaces, which can lead to mold and decay. Also, make sure the air is routed through rigid ducts with as few bends as possible. Flexible pipe might be easier to install, but corrugations and bends resist airflow and decrease the fan's effectiveness.

Choosing a Quiet Fan

Manufacturers use a metric called "sones" to rate the perceived sound level of exhaust fans. The industry describes one sone as the amount of noise a quiet refrigerator makes in a quiet room.

The cheapest fans, the kind that builders and electricians might install if you don't offer any input into the decision, can be rated at 4 sones or higher, which you'll probably find annoying. And if a fan annoys you, you probably won't use it much—especially when you're trying to enjoy a relaxing soak in a tub.

Sone ratings are incremental, like inch measures—2 sones are twice as loud as 1; 3 are three times as loud as 1, etc. A fan rated at 2 sones is pretty quiet. At least two manufacturers, Panasonic® and Broan®, make fans rated at less than 1 sone, some as low as 0.3 sones. At these levels, fans are practically inaudible when running. Larger-capacity fans tend to be noisier than fans intended for smaller spaces. Your best bet is to get the quietest fan that will do the job for the size of the bathroom.

above • Exhaust fans can disappear behind a variety of ceiling light fixtures, such as the one shown here.

right • Because recessed fixtures provide plenty of light, a simple exhaust fan is all that this large bathroom with multiple showerheads needs. Proper fan capacity is based on the room's square footage and the number of fixtures it contains.

Bathroom Fans

CEILING-MOUNTED FANS
$

- Often packaged with a light, ceiling-mounted units include the fan and motor in a metal housing that is recessed into the ceiling.
- Some models draw air through a light fixture with no visible grille.
- Look for a fan with a low sone rating for quiet operation.
- Some can be equipped with infrared heat lamps.
- Small bathrooms need only one ceiling-mounted unit, but larger bathrooms may be better served by two or more.

WALL-MOUNTED FANS
$

- Good solution when framing or accessibility constraints prevent mounting a frame on the ceiling and venting from above.
- Ensure that the unit has a screen to keep critters out.

IN-LINE FANS
$$–$$$

- Fans are installed in the duct itself rather than directly in the ceiling or wall.
- One fan can draw air from multiple rooms, such as adjacent bathrooms, or multiple grilles in a large bathroom.
- More powerful and more expensive than simple ceiling-mounted fans.
- Can be noisy.

CEILING-MOUNTED FANS

WALL-MOUNTED FAN

IN-LINE FAN

left • It's a good idea to locate the exhaust fan as close as possible to the nearest source of vapor—in this case, just over the shower door.

bottom left • Here, two exhaust-fan grilles are installed in the shower enclosure where they can be most effective. The grilles blend inconspicuously into the wood ceiling.

Easy Fan Upgrade

If your old bathroom fan is too noisy and/or not powerful enough, you may be able to purchase an upgrade kit that lets you install a new quieter and more powerful fan into the existing housing. This costs less and is easier to do than replacing the entire unit. Check the manufacturer's website to see if a retrofit is available for your model.

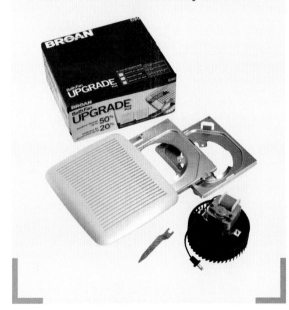

Switches and Timers

The goal is simple: You want the fan to operate whenever someone is in the bathroom and you want it to run for 20 minutes or so after that person leaves to give the fan a chance to clear the room of moisture and odors. There are several ways to accomplish this goal.

Theoretically, a simple on/off switch would do the job, but that assumes everyone who uses the bathroom will turn the fan on when they enter and then come back 20 minutes later to turn it off.

If you are confident that folks will turn the fan on but want to eliminate the inconvenience of returning to the bathroom to turn the fan off, then an inexpensive timer switch will do the trick. Most can be set to run the fan for 10 minutes to an hour.

If you use the bathroom on a rigid schedule—say it's the master bath you use before going to work—you can further automate the process by installing a programmable timer switch that will turn the fan on and off at preset times every day. This type of switch is a good idea if you are using the bathroom fan in lieu of a whole-house fan because you can set it to cycle on and off several times during the day. This is an option that should be discussed with a heating, ventilation, and air conditioning (HVAC) or energy professional because ventilation requirements vary with the size and construction of the house.

To automate the process without a rigid schedule, you could install a motion detector, also called an occupancy detector, in conjunction with a timer. The motion detector will turn the fan on when someone enters the room and then activate the timer when they leave.

This combination unit is a simple way to provide an exhaust fan along with light in the shower.

A very simple timer has presets for running the fan from 10 minutes to an hour.

More complex timers can be programmed to operate at certain times of the day or even to switch on automatically when humidity levels reach a certain point.

Here, a round exhaust fan is set just outside the shower; it complements the round shower light fixture.

Another set-it-and-forget-it approach is to install a humidistat that will turn the fan on when the humidity reaches a preset level and then off when the humidity is reduced to another preset level. A humidistat will protect the bathroom from mildew and other moisture damage, but it doesn't detect odors.

If all of this sounds like a complicated hassle, manufacturers have come up with fans that turn themselves on and off with no switch at all. Panasonic, for example, has a fan that has a motion detector, a humidistat, and a timer built into the unit. This fan can be set to function three ways: It can come on when you enter and turn off when you leave, it can come on when you enter and turn off when the humidity is reduced to a preset level, or it can come on when the humidity reaches a preset level and go off when the humidity is reduced to a lower preset level.

Energy-Saving Ventilation

When energy was cheap, nobody worried much about warm or cooled air leaking out of the house. But in recent decades, with improvements in windows, sealants, and other systems, houses have become increasingly airtight. This greatly reduces energy costs but it creates a new problem—air trapped indoors that is stale or that contains excess moisture, odors, or unhealthy fumes. The challenge is to provide fresh air that is healthy and at the right humidity level while minimizing the energy penalty.

This has lead to the development of heat recovery ventilators (HRV) that pull air out of the house and bring fresh air in. The two streams of air are kept separate, but both streams flow through a heat exchanger and particulate filters. In cold weather, some of the warmth of the heated indoor air is transferred to the incoming outside air. In warm weather, the system works in reverse: Some of the heat from the incoming air is transferred to the outgoing air where it is pulled out of the house.

A similar device called an energy recovery ventilator (ERV) works in the same way, but the heat exchanger is vapor permeable so that moisture can be exchanged between indoors and outdoors. In cold weather, when incoming air tends to be too dry, some of the moisture from the outgoing air can be transferred to the incoming air. In summer, the system can be used in reverse to lower the humidity of incoming air.

ERVs are more expensive than HRVs, and, in general, ERVs are used mostly in climates where it is very hot and humid most of the year. However, climate is not the only factor in choosing between the two. Before making a decision, read up on the subject and talk with a reputable HVAC installer. A good source of information on this and other indoor air-quality issues is Green Building Advisor (greenbuilding advisor.com).

Supplemental Heaters

You might be perfectly comfortable setting the whole-house thermostat at 65°F on a cold winter day as long as you can pull on a sweater to hang out in the living room or toss an extra blanket on the bed at night. But 65°F will seem mighty chilly when stepping out of a bath or shower. Adding auxiliary heat to the bathroom can make you more comfortable while saving energy dollars by allowing you to add warmth to the bathroom only when you need it.

The type of auxiliary heating you choose will depend to some extent on the type of heating system you have for the rest of the house. If it is a hydronic system (that is, it runs on hot water), you may be able to hook up a toe-kick heater beneath the cabinet, a heated towel bar, or even a low-profile European-style radiator mounted on the wall.

If you will be doing an extensive remodel that includes replacing the bathroom floor or the ceiling below the bathroom, consider installing radiant floor heat. This provides a very pleasant heat that feels great under bare feet. And with no visible parts, in-floor heat saves space and won't interrupt your décor.

The most common way to provide radiant floor heat is through hot-water tubing, but this is usually done as a main source of heat for a whole house. If the rest of the house is heated with hot water and you are installing a new floor, you may be able to retrofit hot-water tubing under the bathroom floor. However, in most cases it makes sense to retrofit electric radiant floor heat.

Electric radiant heating is most typically used as a supplemental heat source, so it's perfect for adding that extra warmth to the bathroom. It consists of electric resistance cables that are often preattached to a mat to make installation easier. The mats are installed over the subfloor.

If you are gutting the bathroom walls and ceiling but preserving the floor, there are radiant panels designed to be installed under drywall. If you don't want to pull up the existing bathroom floor, another alternative is electric radiant heat pads designed to be installed between joists under the subfloor.

There are other types of electric resistance heaters including heaters combined with ceiling vent fans and fixtures that accept infrared bulbs. However, because heat rises, these are not the most efficient alternatives. You also can install electric radiant heating in the form of a baseboard or unit mounted on the wall. Or, to save space, you can get a unit sized to fit between wall studs.

Another approach, if you have the space and budget, is to treat the supplemental heat source as a luxurious focal point by installing a gas fireplace in the bathroom.

A good gas fireplace will take a bite out of your construction budget, but what could be cozier tucked between the tub and shower?

You can buy premade towel warmers that employ hot water or electricity. Or you can create a custom hot-water towel rack like this one that makes a bold statement and is big enough to contribute significant heat to the room.

Heaters

CEILING-MOUNTED HEAT LAMP
$

- Inexpensive and unobtrusive.
- Most typically used outside a shower or tub.
- Heat lamps can be incorporated with exhaust fans.
- Use with timer to limit run time.
- Heats a limited area.

TOE-KICK HEATER
$-$$

- Available in electric or hydronic models.
- Makes good use of wasted space under a cabinet.
- Fan distributes warm air.
- Depending on location of hot-water pipes or electric cables, may be possible to add without major demolition.

WALL AND CEILING HEAT PANELS
$-$$

- Electric radiant panels available in a number of sizes that run on 120- or 220-volt current.
- Installed behind drywall, panels are out of sight and take up no floor space.
- Can be connected to a programmable thermostat to provide scheduled heat.

ELECTRIC RADIANT FLOOR MAT
$-$$

- Grid of wires, usually attached to a mat, can be installed under tile or floating floor or under subfloor between ceiling joists.
- Standard and custom sizes available.
- Available for wet locations such as the shower.
- Out of sight; doesn't take up any floor area.
- Adds little floor height so it's suitable for a remodel.

TOWEL WARMER/RADIATOR
$$

- Electric and hydronic models are available.
- Depending on heating loads, can be used for supplemental heat or as a sole heat source for bathroom.
- Keeps towels warm.
- Requires no floor space.

RADIANT FLOOR
$$-$$$

- Installation requires removal of floor or access to bathroom floor from below.
- Available as hydronic or electric.
- Provides even, gentle heat.
- Takes up no floor space.

EUROPEAN-STYLE RADIATOR
$$$

- Low profile; takes up less space than a conventional radiator.
- Wall-mounted models take up no floor space.
- Can be custom-made to fit space, including curves.
- Can replace conventional radiators without the need to overhaul existing hydronic heating system.

High Style and High Efficiency

You need only glance at the photos to see that this master bathroom is quite original in style. Less obvious are the innovations that make this space a highly efficient energy miser. The bath is part of a house designed for net zero energy use—the goal is for the house to use no more energy than is produced on the site.

Efficient use of energy begins with efficient use of space—smaller spaces simply use less energy, notes architect Marc Sloot of SALA Architects, who designed this home in a suburban neighborhood near Minneapolis. For this reason, the bathroom, and indeed the whole house, packs a lot of function into a compact footprint. For example, the Japanese-style soaker tub is designed to be used sitting up rather than reclining as you would in a Western-style tub. It has a seat and it's deep enough to sit submerged to your neck, while taking up much less floor space than a conventional tub.

The tub shares a wall with a glass-enclosed shower. The enclosure doesn't interrupt the horizontal flow of the wall tile, contributing to the open feel of the space. The soffit bows out to follow the contour of the tub. In addition to contributing to the visual flow of the space, the soffit has a practical purpose, Sloot explains. Wherever there is a recessed ceiling fixture in the house it is always in a soffit. This way, fixtures never pierce the ceiling itself, which would create air and moisture leaks.

For minimum energy use and maximum life, the shower light is LED. The vanity mirrors are lit by narrow LED strips at the sides, a sleek look that conserves space and provides even, shadow-free light that's perfect for shaving or applying makeup.

Two important energy-saving features are completely hidden. The entire house, including the master bath, is heated by efficient hydronic radiant heat in the floor. And the house uses an energy recovery ventilation system that removes moisture and odors while capturing about two-thirds of the heat from the exhausted air (see "Energy-Saving Ventilation," p. 149).

above · The tub surround sweeps back to make room for steps that would otherwise encroach on the tight floor space. In addition to saving space, the steps add a sculptural element.

facing page top · This bathroom has a style all its own while employing many energy-saving features.

facing page bottom left · Simple-looking vanity faucets from Brizo can be turned on and off with a touch. To conserve water, they have a timer that automatically shuts off the water when a motion detector senses no one is present.

facing page bottom right · All of the cabinetry and the soffit face are constructed of panels faced with Echo Wood veneer, which consists of reconstituted wood fibers made to look like hardwood—in this case, cherry. Use of this product conserves hardwood forest around the world. In this bath, the grain is run horizontally to make the space feel more expansive.

STORAGE

From bathing to grooming, the bathroom is a place of varied activities,

each requiring its own set of items. To avoid overwhelming the bathroom with clutter,

you'll need a storage strategy that may include built-ins, open shelving,

freestanding furniture, and, of course, a medicine cabinet.

One consequence of our quest for a spa experience in the bathroom is the need for more places to store stuff. A small powder room may need nothing more than a medicine cabinet over the sink and a place to store toilet paper and hand towels in the vanity or wall cabinet. A master bath, on the other hand, may need storage for bath towels, shampoos, lotions, bubble bath, razors, medicines, toothbrushes, and a whole lot more—sometimes even a television and/or music system.

In designing your bathroom storage, it's a good idea to make a list of all the things you think you'll store in the bathroom—start by writing down everything that's crammed into the bathroom you use now. Then add the things you don't currently have on hand in the bathroom but wish you did. Now divide the stuff into two categories: things you use every day and things you use only occasionally. Everyday items, such as toothbrushes, toothpaste, and soap, should be easily reached without bending down or reaching up. That means out on the vanity counter or nearby shelf or, if it's an item you want out of sight, in a drawer just below the counter.

Inside a vanity cabinet, which is partially taken up with plumbing pipes, is often the most awkward-to-reach storage and is best reserved for items you'll use less often, such as extra toilet paper or bottles of shampoo. A vanity with open shelving below can be easier to access than one with doors.

Open shelves can be a beautiful design element and a great place to store items you want to display, such as perfumes, candles, or a jar of decorative soaps or bath beads.

Most bathrooms can benefit from a floor-to-ceiling cabinet for storing bulkier items like bath towels. If the bathroom footprint is tight, it might be possible to steal storage space from an adjacent room with a door or doors that open into the bathroom. If not, storing towels in a linen closet near the bathroom is not a bad compromise.

And don't overlook the storage potential of the wall behind the toilet. You can buy a freestanding unit with legs that straddle the toilet, or you can mount shelves or a cabinet behind.

A shallow cabinet with drawers, built to fit between two wall studs, puts hand towels and toiletries right where you need them without taking up space in the bathroom.

above • Cabinets and shelves flanking this tub create visual interest along with lots of storage.

facing page bottom • The owners of this bath took advantage of a high ceiling to add storage that is accessible by a library-style ladder that runs on a track.

Cabinetry

From Colonial to postmodern, whatever the architectural style of your home or the motif of your bathroom, you'll have no problem finding premade cabinets to match. And while cabinets designed to look built in are still very popular, a recent trend is to install cabinets built to look like freestanding pieces of furniture. You might even adapt a table or buffet into a new life as a bathroom vanity. (See "Built-In and Wall-Mounted Cabinets," p. 168.)

Cabinets fall into three broad categories: stock, semi-custom, and custom. Semi-custom generally costs more than stock, and custom costs more than semi-custom. However, price and quality can vary greatly within each category, so use these industry labels only as a starting point. For bathroom cabinets, stay away from particleboard unless it is specifically designated for use in wet areas. Regular particleboard will swell if it gets wet. Look for solid plywood construction with strong joinery and with doors of plywood or solid wood. If the cabinets have drawers, look for interlocking joints—dovetails are traditional but there are other types. Avoid drawers made of thin material joined with staples.

A combination of shelves for things you want to display and drawers for things you want to stash is a great formula for taming clutter. There's no room for drawers in front of inset sinks, but these are handy spots to mount towel racks.

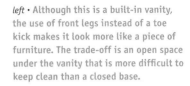

left • Although this is a built-in vanity, the use of front legs instead of a toe kick makes it look more like a piece of furniture. The trade-off is an open space under the vanity that is more difficult to keep clean than a closed base.

bottom left • A tall cabinet makes optimum use of space. This bold black example also divides the sink area from the makeup station.

bottom right • The rich red cherry wood of this tall Shaker-style cabinet makes it a focal point against the light tones of the bathroom. The cabinet appears freestanding, but the slots in the base are a clue that it was custom-made to fit the space. The slots allow heat to escape from a baseboard heater behind the cabinet or perhaps a kick-heater below.

STOCK

Stock cabinets are the most economical and are available in limited sizes, colors, and finishes. They're kept in stock in big-box stores and kitchen and furniture showrooms. Many stock cabinets are made with particleboard, but you'll find good-quality plywood cabinets too. And by intelligently adapting the design and layout of your bathroom to accommodate good-quality stock cabinets, you can create a gorgeous bathroom while perhaps leaving money in the budget for other touches, such as a glass or stone custom countertop or that high-end tile you're craving.

SEMI-STOCK

These are really stock cabinets with more options—just like stock, the boxes come in 3-in. increments, so semi-stock doesn't allow any more flexibility in designing the bathroom space. But you can order them with your choice of door styles, pulls and knobs, wood species, and finishes. Also, it's easier to find semi-stock cabinets made of good-quality plywood and higher-quality hardware, such as full-extension, ball-bearing drawer slides rather than simple epoxy-coated slides. You will have to order the cabinets, and lead times vary by manufacturer.

above • A freestanding vanity is easy to install and provides an old-fashioned feel. Although probably not a big issue in a lightly used powder room, cleaning is complicated by tough-to-reach narrow spaces at the sides and bottom.

facing page top left • The texture of the rough-sawn boards used to custom-build this vanity is a nice counterpoint to the thick and smooth concrete counter.

facing page top right • Vanities don't always need to be placed against a wall. Here, an octagonal version with two sinks separated by a two-sided mirror is the centerpiece of a his-and-hers master bath. While you won't find an octagonal stock or semi-stock cabinet, you could use stock to create a rectangular vanity island.

facing page bottom • Wall-mounted vanities are most often used in bathrooms with a contemporary style. The uninterrupted floor evokes a sense of spaciousness and is easy to clean.

CUSTOM

Got your heart set on cabinets of bird's-eye maple, bubinga, or some other unusual wood? Have a certain type of inlay in your mind's eye? A skilled cabinetmaker can create custom cabinets to take your vision as far as your budget is willing to go. Custom cabinets can be built to fit your space exactly, avoiding the need for the awkward filler panels usually necessary with stock and semi-stock cabinets. You can even have cabinets made to fit a curved wall.

Depending on what you order and from whom you are buying, lead times can be long—often months—so that's something to ask about when shopping around. There are a few companies that specialize in custom cabinets, but you might do just as well or better working with a local shop or even an individual cabinetmaker in your area. You can visit the cabinetmaker's shop, see samples of his or her work, and talk with people in your area who have used that cabinetmaker's services. Collaborating with a cabinetmaker and your bathroom designer—if you are using one—can be a creative and satisfying experience.

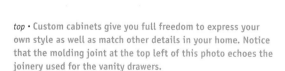

top • Custom cabinets give you full freedom to express your own style as well as match other details in your home. Notice that the molding joint at the top left of this photo echoes the joinery used for the vanity drawers.

right • A few special cuts on the vanity apron along with the mirrors above give this bathroom an Asian flair.

above • Only custom cabinetry can achieve the rustic handmade look that complements this cottage-style bathroom.

left • By slicing a board through its thickness, opening the two pieces like a book, and gluing the pieces edge to edge, custom cabinetmakers can create "book-matched" panels such as those used in this vanity's doors and draw fronts.

Cabinet Types

Although spending more usually buys better materials and finishes and more careful assembly, that's not always the case. Compare features, construction, fit, and finish carefully.

STOCK
$

- Most economical choice.
- Made in 3-in. increments, so filler panels are often needed.
- Limited number of styles and finishes.
- Kept in stock at big-box stores and kitchen-and-bath showrooms.
- Stock cabinets can be a good buy, but check that materials are water resistant and robust enough for everyday use. Doors and drawers should operate smoothly.

SEMI-CUSTOM
$$

- More choices than stock for finishes, materials, styles, wood types, and hardware.
- Manufactured in 3-in. increments.
- Can be ordered at home centers or kitchen-and-bath showrooms. Lead time for delivery varies by manufacturer.
- Materials and construction techniques typically are of higher quality than stock cabinets. Look for drawers with dovetails or other interlocking joints, full-extension drawer slides, sturdy plywood cabinet boxes, and hardwood drawer sides and face frames.
- When ordering cabinets through a kitchen-and-bath showroom, ask whether their services include a personal visit by a designer to ensure cabinets are correctly sized and properly installed.

CUSTOM
$$$

- Can be built to specified height, width, and depth.
- Materials and construction typically are of the highest quality.
- You can specify wood species, trim details, and hardware, including door and drawer pulls and hinge type.
- Molding profiles can exactly match the rest of the room, creating seamless-looking built ins.

- Lead times can be quite long.
- Available from manufacturers that specialize in custom cabinetry, such as Rutt®, and ordered through local retailers. Working directly with a local cabinet shop or individual cabinetmaker is an attractive option.
- Custom makers offer highly personalized design services.

facing page · Simple flat panels show off beautiful wood in these cabinets that are punctuated with towel cubbies painted to match the walls.

bottom left · In this unusual vanity, the raised double sink, counter, and side are all one cream-colored piece that harmonizes with the wood shelves and drawers below.

below · Bold blue cabinets and ceiling work with paler blue walls and white trim to create a vibrant and handsome bathroom.

Details Make the Difference

Whether as an aesthetic touch, such as a special drawer joint, or a clever solution to a practical problem, such as where to stash a child's stool, thoughtful details can make your bathroom a unique and special place.

1. Typically, the sink drainpipe precludes the use of a drawer under the sink. Here the drawer is configured to literally get around that problem. 2. The stools little kids need to reach the sink can get underfoot, especially in a small bathroom. Here, full-extension drawer slides allow stools to slide neatly under the sink vanity. 3. Well-made dovetail joints and solid wood construction are hallmarks of the best cabinetry. For an extra touch of elegance, this drawer front meets the side of the vanity with a miter. 4. Because the toilet paper holder is always near one of everybody's favorite reading spots, why not incorporate a shelf for books? 5. Manufacturers are adapting kitchen cabinet pullouts for use in the bathroom. The narrower shelf that would hold spices in the kitchen is perfect for makeup and small bottles of lotion, while the wider pullout stashes the hair dryer and other larger items.

Built-In and Wall-Mounted Cabinets

As with kitchens, the most common approach to installing bathroom cabinets is to run them into corners, screw them in place, and then install a counter, and usually backsplashes, to fit. This built-in approach is undoubtedly the most efficient use of space—something that is often at a premium in a bathroom. And backsplashes protect walls, keep things from falling behind cabinets, and simplify cleaning.

Built-in cabinets are usually set on the floor with a toe kick, but there are also cabinets designed to be mounted to the wall with space left below the cabinets. This sacrifices a bit of storage space but leaves an uninterrupted floor that makes the room appear larger and makes the floor easier to clean. Another advantage of a wall-mounted cabinet, especially a vanity, is that you can mount it at any height you like. Stock vanities are set at a standard height designed to be comfortable for people of average height. If the main user or users of a bathroom is significantly shorter or taller than average, a wall-mounted vanity can accommodate them.

Because its full weight is supported by framing inside the wall, a wall-mounted vanity or other cabinet requires more careful planning so blocking and structural wall framing can be correctly located.

above • Although this looks like three pieces of freestanding furniture, the sink "table" has no back legs and is mounted to the wall, as are the two chests of drawers flanking it.

facing page top left • These elegant semi-stock cabinets include a vanity and narrow cabinets flanking a mirror. The narrow filler strip at the wall is barely noticeable, making the cabinets look as if they were custom-made to fit.

facing page top right • A corner cabinet is a space-saver in any room. This built-in version creates a display case behind the tub in space that might otherwise be wasted.

facing page bottom • With a makeup station flanked by his-and-hers vanities, this cabinet array offers lots of functionality with natural light from an arch-top window.

Opening Up a Dark, Narrow Space

Renovation is often about doing the best you can with an existing space. Here, the existing bathroom was long and narrow, a fact that was emphasized because a linen cabinet, vanity, and tub were lined up on one wall with partition walls between them. The partitions prevented natural light from a window from reaching beyond the tub and toilet area. They also created a narrow corridor so that when you opened the bathroom door your eye was drawn through a narrow space to the toilet on the back wall.

The renovation goal was to create a fun-looking, light-filled bathroom to serve the needs of two young boys and occasional guests. So architect Sophie Piesse started by removing all the partitions, including the linen cabinet, to open up the space. The new tub now occupies the back wall. The toilet is next to the tub, hidden and separated from the vanity by a short partition wall. The long vanity has a long sink with two faucets, so the boys can get ready for school at the same time.

It seems counterintuitive, but one way to make a narrow space seem more spacious is to emphasize horizontal lines. The eye is drawn along the room by the long vanity topped by a long mirror, itself topped by a continuous LED light strip. The empty space between floor and vanity creates another strong horizontal line, and the horizontal grain of the bamboo vanity drawer fronts also contribute to this effect.

The choice of bamboo was practical as well as aesthetic, Piesse notes. "Bamboo is less likely to be damaged by moisture, an important consideration for a boy's bath that is likely to get steamy. And all those drawers in the vanity provide the storage that was lost by eliminating the linen closet.

"I do a lot of drawers," Piesse noted. "People don't get a lot of storage behind doors."

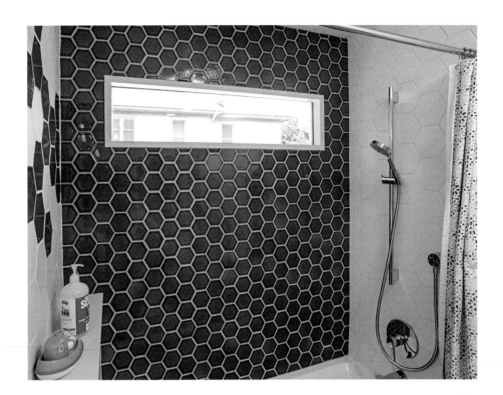

above · The strong horizontal lines of the vanity, mirror, and light help make this narrow bathroom seem more spacious.

top right · This high horizontal window provides light and a view without sacrificing privacy. A local glass shop made the double-pane glass that is surrounded by quartz synthetic countertop material that is impervious to water. The glass is held in place by exterior stops, so if ever necessary, the glass can be removed from the outside without disturbing the interior frame or tile.

left · The same synthetic quartz was used for the vanity top and window frame as well as to create a nook for the toilet. The nook provides a shelf that extends across the end of the bathtub, conveniently using up a few inches of wall left by a standard-size tub.

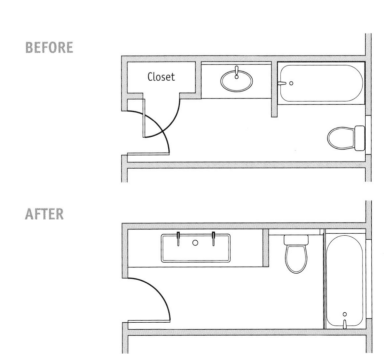

BEFORE

AFTER

Stand-Alone Furnishings

Adding freestanding furniture can make the bathroom more inviting and comfortable while helping the space blend more seamlessly with the décor of your home. This can be as simple as bringing in a favorite side table for toiletries or perhaps a trunk to store towels.

You also can purchase freestanding cabinets designed specifically for bathrooms, including freestanding vanities. Or you might be able to adapt an existing table, dining room buffet, or chest of drawers into a one-of-a-kind vanity.

Tables work well to support vessel sinks. For one thing, a table with a vessel sitting on it just looks natural. Beyond that, standard table height is too low for an inserted sink, but it can be just right for a vessel. And a vessel requires a smaller hole in the top, which is less likely to weaken the table.

You might find a chest of drawers that is the right height for an inserted sink, but that might require removal of internal drawer dividers and frames to make room for plumbing. Any good cabinet shop should be able to assess whether a given piece of furniture can be adapted into a vanity. Of course, you can always commission a cabinet shop to create a freestanding vanity or other bathroom cabinet from scratch.

Freestanding tables tend to pair nicely with vessel sinks. In this powder room, a table with a wooden top is a perfect home for a wooden vessel sink.

Old house doesn't dictate old style. Though quite modern in appearance, this freestanding vanity picks up on the massiveness of the house's old beams and the color of the woodwork.

A simple lightweight shelving unit keeps towels handy next to the shower.

A freestanding cabinet placed strategically between a pedestal sink and the shower has an open shelf for bath towels and a closed cabinet for toiletries.

Medicine Cabinets

A medicine cabinet seems like standard equipment in a bathroom, and there's a good reason they are so ubiquitous. It's simply handy to have some extra storage hidden behind a mirror and within easy reach to stash away all the odds and ends that collect around a sink. It's a practical necessity for a bathroom with a pedestal sink, which offers very little counter space for toiletries and no storage below.

If your bathroom doesn't already have a medicine cabinet, adding one can be as easy as screwing a wall-mounted model to the wall. And installing a recessed cabinet is not a daunting task as long as the wall is not load-bearing and there is no wiring or plumbing in the way. Medicine cabinets are usually about 30 in. wide to fit in two 16-in. on-center stud bays, so part of the stud separating the bays needs to be cut out—redistributing that stud's load in a finished load-bearing wall makes the job more tricky.

Of course, if your bathroom is part of new construction or the walls are being stripped to the studs, you can use any medicine cabinet you like—stud framing can be designed to fit around it. You can also have a recessed cabinet custom-made to fit any situation. Trim surrounding the cabinet can be elaborate, or it can disappear altogether in favor of an unadorned mirror, making this kind of installation very flexible.

If you are looking for a cabinet to match the architectural style of your home, a variety of online sources offer medicine chests to match any décor. Want something funky or unique? Check out Etsy.com—you'll find hundreds of handmade, one-of-a-kind choices.

A pair of medicine cabinets elegantly framed in the wall harmonize visually with the doors of the vanity below.

Placing a medicine cabinet with a mirror to the side of the main mirror makes it easy to see the side and back of your head.

left • A classic wall-hung medicine cabinet looks appropriate in a period-style bathroom with wainscoting and pedestal sink. And because the cabinet and sink occupy a niche, a wall-hung cabinet is more accessible than a recessed cabinet would be.

above • Although this wall-mounted medicine cabinet veers from the old-style furnishings of this timber-frame house, it's simple enough to fit in just fine.

left • Freed from mirror duty by a large mirror in front of the vanity, this recessed medicine cabinet unobtrusively provides handy storage.

Open Shelves and Cubbies

Whether formed from simple boards or adorned with elaborate moldings, open shelves store things you want to display. For example, instead of stashing bath towels in an enclosed cabinet, perhaps you'd like to show off towels in colors selected to complement your wall color—and guests won't have to ask where they are. Glass jars filled with special soaps or bath salts are other candidates for handy display.

Custom-built cabinets get expensive and impractical for small spaces, while shelves are a much more flexible and inexpensive way to turn nooks and awkward spaces into storage—a small strip of wall next to a shower, for example, could become the perfect spot for washcloths and extra shampoo bottles.

1. Short glass shelves utilize an awkward space under a sloped roof. 2. Shelves of various sizes along with a stack of drawers take advantage of all the space around the tub in this old-fashioned bathroom. 3. These open shelves are conveniently located next to the tub. Their smooth brown wood contrasts with the texture and color of the wall they meet. 4. Simple wooden shelves are right at home in this contemporary bathroom. 5. The plumbing chase behind a showerhead often offers the opportunity to add a column of narrow shelves that are handy for grabbing a towel as you step out of the shower. This one has a cabinet below. 6. This little powder room is located next to a staircase. The owners grabbed some of the space under the stairs to create a mini-library.

Reconfiguring a Master Suite for Today's Lifestyle

When this house was built in the 1950s, no one would have thought to devote much space for a bathroom. After all, it wasn't a place people lingered, and how much room did you need for a toilet, sink, and tub/shower combo? And so, the master suite of this home in a suburb of Philadelphia had a cavernous master bedroom with a cramped master bath and a tight master closet.

"The space was really vast but not comfortable and not intimate," said Architect Jeff Krieger of Krieger + Associates Architects, whose plan called for removing all the existing partition walls and reconfiguring the space to create a spacious new bath, a commodious closet, and an invitingly intimate bedroom. There was even room to add a little office niche. The new bath features a two-person shower, a separate soaking tub, two sink vanities, and a private nook for the toilet.

Bathrooms of the 1950s tended to have tile that was pink and black or avocado and mustard, Krieger noted, so the owners had no interest in restoring the bathroom to period style. Rather, they wanted a look that is classic and timeless. The old-style soaking tub, chrome fixtures, beadboard wainscot, and basketweave floor tile evoke the 1920s. Meanwhile, the glass shower enclosure is a thoroughly contemporary element that allows natural light to fill the space and makes the bathroom seem more open and spacious.

In addition to doors into the bath and the closet, the new configuration has a door between those two spaces. The ability to move between the bath and the closet greatly improves the traffic flow within the master suite. There was no room for a swinging door between the bath and closet, so a sliding barn-style door was used.

"The barn door added a lot of character to both rooms," Krieger said. "It's open most of the time, but when the owner has guests over who will use the bath she doesn't want to have to tidy up the closet, so she can just close the door."

The controls for the two overhead rain showerheads are just inside the door. This nice touch allows the user to adjust the water before getting under the showerheads. For an alternative shower experience, the wall-mounted showerhead is height adjustable or can be used handheld.

Twin vanities flank the entrance to the bathroom. Their dark color makes them a focal point against the light tones used elsewhere. The large mirrors bounce light from the opposing windows, increasing the natural light in the space.

A decidedly contemporary glass-enclosed shower blends seamlessly with an old-fashioned tub, fixtures, and floor tiles, proving that you can enjoy modern amenities in a classic setting.

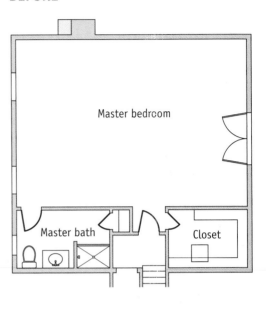

Master bedroom

Master bath

Closet

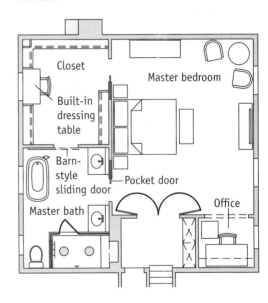

Closet

Master bedroom

Built-in dressing table

Barn-style sliding door

Pocket door

Master bath

Office

The large, sliding barn-style door between bath and closet improves traffic flow without getting in the way as a swinging door would.

RESOURCES

ORGANIZATIONS AND ASSOCIATIONS

The American Institute of Architects
www.aia.org
(800) AIA-3837

Fine Homebuilding magazine
www.finehomebuilding.com
(800) 309-8919

Green Building Advisor
www.greenbuildingadvisor.com
(800) 943-0253

MIA +BSI: Natural Stone Institute
(formerly Marble Institute of America
and Building Stone Institute)
www.marble-institute.com
(440) 250-9222

National Association of
Home Builders
www.nahb.org
(800) 368-5242

National Association of the
Remodeling Industry
www.nari.org
(847) 298-9200

National Kitchen & Bath Association
www.nkba.org
(800) 843-6522

The RL Mace Universal Design
Institute
www.udinstitute.org
(919) 608-1812

Tile Council of North America
www.tcnatile.com
(864) 646-8453

MANUFACTURERS AND SUPPLIERS

American Standard
www.americanstandard-us.com
(800) 442-1902

Armstrong
www.armstrong.com
(800) 233-3823

Brizo
www.brizo.com
(877) 345-2749

Broan-Nutone
www.broan-nutone.com
(800) 558-1711

Concrete Network
www.concretenetwork.com
(866) 380-7754

Delta Faucet Co.
www.deltafaucet.com
(800) 345-3358

Kohler Co.
www.kohler.com
(800) 456-4537

Mannington
www.mannington.com
(800) 356-6787

MTI Baths
www.mtibaths.com
(800) 783-8827

PaperStone
www.paperstoneproducts.net
(360) 538-1480

Philip Watts Design
www.philipwattsdesign.com
(44) 115-926-9756

Rev-a-Shelf
www.rev-a-shelf.com
(800) 626-1126

CREDITS

p. 17: Jim Westphalen Photography/Collinstock, design: Birdseye Design/Build, Wagner Hodgson Landscape Architecture
p. 18: Arkin Tilt Architects
p. 19: Arkin Tilt Architects (left and right)
p. 20: Andrea Rugg Photography/Collinstock, design: Sylvestre Remodeling & Design (top)
pp. 20–21: Brian Vanden Brink, design: Phi Home Design
p. 21: Andrea Rugg Photography/Collinstock, design: Sylvestre Remodeling & Design (top)
pp. 22–23: Lana Barbarossa, Allied ASID, designer, Andrea Rugg photography/Collinstock
p. 22: courtesy Delta Faucet Co. ©2018 (bottom)
p. 23: courtesy Delta Faucet Co. ©2018 (bottom)
p. 24: Randy O'Rourke (top); Lana Barbarossa, Allied ASID, designer, Andrea Rugg photography/Collinstock (bottom)
p. 25: Lana Barbarossa, Allied ASID, designer, Andrea Rugg photography/Collinstock (top and bottom)
p. 26: Greg Premru Photography
p. 27: Greg Premru Photography (top and bottom)

CHAPTER 2
p. 28: ©CarolynBates.com, interior design: Cecilia Redmond, LEED AP, Allied ASID, Redmond Interior Design, architects: Byron W. Haynes and R. Andrew Garthwaite, Haynes & Garthwaite Architects
p. 30 Brian Vanden Brink, design: Hutker Architects (top); Eric Roth, Butz+Klug Architects, Boston (bottom)
p. 31: Brian Vanden Brink, design: Eric A Chase Architects
p. 32: Lana Barbarossa, Allied ASID, designer, Andrea Rugg photography/Collinstock (left)
pp. 32–33: Udine Prohl, design: Saunders Architecture
p. 33: Andrea Rugg Photography/Collinstock, design: CF Design Ltd. (right)
p. 34: Hulya Kolabas, design: Lorraine Bonaventura Interior Design (top); ©CarolynBates.com, interior design: Diane Gabriel, artist designer (bottom)
p. 35: courtesy Kohler Co.
p. 36: Brian Vanden Brink, design: Carl Solander, Architect
p. 37: Randy O'Rourke
p. 38 Brian Vanden Brink, design: Eric A Chase Architects (top); Randy O'Rourke (bottom)

p. 39: Eric Roth, Butz+Klug Architects, Boston (top left); courtesy Kohler Co. (bottom left); Andrea Rugg Photography/Collinstock, design: U + B Architecture & Design (right)
p. 40: Brizo® Kitchen and Bath Co.
p. 41: Brizo® Kitchen and Bath Co. (top left); courtesy Kohler Co. (top right); Brizo® Kitchen and Bath Co. (bottom left); courtesy Kohler Co. (bottom right)
p. 42: Randy O'Rourke
p. 43: Randy O'Rourke (top left and top right)
p. 44: Brian Vanden Brink, design: Blas Bruno Architecture
p. 45 Randy O'Rourke (top); Andrea Rugg Photography/Collinstock, design: U + B Architecture & Design (bottom left); Brian Vanden Brink, design: Catalano Architects (bottom right)
p. 46 Randy O'Rourke (top)
pp. 46–47: Susan Teare, general contractor: Donald P. Blake Jr, Inc., architectural and interior design: Cushman Design Group
p. 47: Photography by Troy Theis, design: Martha O'Hara Interiors (top)
p. 48: courtesy Kohler Co. (top); courtesy Kohler Co. (bottom)
p. 49: Eric Roth, design: ZEN Associates, Boston
p. 50: courtesy MTI Baths
p. 51: Jim Westphalen Photography/Collinstock, design: Elizabeth Herrmann Architecture + Design
p. 52 DXV (top); Ryann Ford Photography (bottom)
p. 53 Eric Roth, design: Butz+Klug Architects, Boston (top); Jim Westphalen Photography/Collinstock, design: Redmond Interior Design (bottom)
pp. 54–55: Randy O'Rourke
p. 54: Randy O'Rourke (bottom)
p. 55: Randy O'Rourke (top right and bottom)

CHAPTER 3
p. 56: Randy O'Rourke
p. 58: courtesy Kohler Co. (top); Brian Vanden Brink, design: Sally Weston Architect (bottom)
p. 59: Ryann Ford Photography, design: Poteet Architects (left); Eric Roth, design: Thomas Buckborough Design, Boston (right)
p. 60: ©CarolynBates.com, interior design: Amy Thebault Design (top); Michael Moran/OTTO (bottom)
p. 61: Eric Roth (top); courtesy Kohler Co. (bottom left); courtesy Kohler Co. (bottom right)

p. 62: ©CarolynBates.com, interior design: Bob and Aida Luce (left); photography by Troy Theis, designed by Martha O'Hara Interiors (top right); courtesy American Standard (bottom right)
p. 63: Brian Vanden Brink (top left); Eric Roth, design: Butz+Klug Architects, Boston (top right); Randy O'Rourke (bottom left); Randy O'Rourke (bottom right)
p. 64: Brian Vanden Brink, design: Joe Adams, Architect (left); Randy O'Rouke, Frank Shirley Architects (top right); Susan Teare, design: Peregrine Design/Build, Stephen Alastair Wanta Architect (bottom right)
p. 65: Brian Vanden Brink, design: Polhemus Savery DaSilva Architects Builders
p. 66: courtesy Kohler Co. (top); Eric Roth, design: Michael Collins Architects, Boston (bottom left); Hulya Kolabas (bottom center); Susan Teare, design: Peregrine Design/Build (bottom right)
p. 67: Chipper Hatter, design: Shiny Bones, Myca Loar—designer (top); Eric Roth
p. 68: Brian Vanden Brink, design: Carriage House Studio Architects (top); ©CarolynBates.com, architect: Elizabeth Brody, artist: Linda Provost (bottom left); Brian Vanden Brink, design: Polhemus Savery DaSilva Architects Builders (bottom right)
p. 69: Photography by Troy Theis, design: Acacia Architects, builder: Ben Quie & Sons Builders (left); courtesy American Standard (right)
p. 70: Randy O'Rourke
p. 71: Randy O'Rourke (top, bottom left, and bottom right)
p. 72: Brian Vanden Brink, design: Dominic Mercdante Architect
pp. 72–73: Mark Lohman, design: Caroline Burke Designs
p. 74: Brian Vanden Brink, design: Dominic Mercdante Architect (top); Undine Prohl, Safdie Rabines Architects (bottom)
p. 75: Brian Vanden Brink, design: South Mountain Company
p. 76: Brian Vanden Brink, design: Hutker Architects (top); Chipper Hatter, design: Countertop Shoppe, Lynda Fisher—designer (bottom)
p. 77: Chipper Hatter, Shiny Bones, Myca Loar—designer (top); courtesy PaperStone (manufactured by Paneltech International) (bottom)
p. 78: Randy O'Rourke

p. 79: photography by Troy Theis, design: Meghan Kell Cornell, AIA, and Katherine Hillbrand, AIA, of SALA Architects, Inc.
p. 80: courtesy American Standard (top); courtesy Kohler Co. (bottom)
p. 81: courtesy Kohler Co. (top and bottom)
p. 82: Brian Vanden Brink, design: Howell Custom Building Group (top); Randy O'Rourke (bottom)
p. 83: Brian Vanden Brink, design: Dominic Mercdante Architect (top left); ©CarolynBates.com, architect and interior design: Nils Luderowski (top right); Randy O'Rourke (bottom)
p. 84: courtesy Kohler Co. (top); courtesy American Standard (bottom)
pp. 84–85: courtesy DXV
p. 85: courtesy Kohler Co. (top and bottom right)
p. 86: courtesy DXV (top)
pp. 86–87: courtesy Kohler Co.
p. 87: courtesy Kohler Co. (top)
p. 88: courtesy Kohler Co (top); courtesy DXV (bottom)
p. 89: Karen Melvin (top), design: Pale Ale by Philip Watts Design (bottom)
p. 90: Randy O'Rourke
p. 91: Randy O'Rourke (top left, top right, bottom left, and bottom right)

CHAPTER 4
p. 92: Andrea Rugg Photography/Collinstock, design: Otogawa-Anschel Design + Build
p. 94: Andrea Rugg Photography/Collinstock, design: Otogawa-Anschel Design + Build
p. 95: Brian Vanden Brink, design: Jackie Silverio Architect (top left); Andrea Rugg Photography/Collinstock, design: Otogawa-Anschel Design + Build (top right); Randy O'Rourke (bottom)
pp. 96–97: Photo and concrete installation by Colorado Hardscapes
p. 97: Andrea Rugg Photography/Collinstock (top); Randy O'Rourke (bottom)
p. 98: Eric Roth
p. 99: Chipper Hatter, design: Roomscapes, Debbie Nassetta—designer
p. 100: Eric Roth, design: Trikeenan Tileworks, Swanzee, NH
p. 101: ©CarolynBates.com, interior design and contractor: Sandy Lawton, ArroDesign (top); Mark Lohman, design: Caroline Burke Designs (bottom)
pp. 102–103: ©CarolynBates.com, interior design: Cecilia Redmond,

LEED AP, Allied ASID, Redmond Interior Design, architects: Byron W. Haynes and R. Andrew Garthwaite, Haynes & Garthwaite Architects

p. 102: Brian Vanden Brink, design: Knickerbocker Group, Interior design: Urban Dwellings (bottom)

p. 103: ©CarolynBates.com, architect: Matthew Milnamow, AIA, LAN Associates (bottom left); Hulya Kobalas, Michele Hogue Interior Design (right)

p. 104: Andrea Rugg Photography/ Collinstock, design: Otogawa-Anschel Design + Build (top); Randy O'Rourke (bottom)

p. 105: Randy O'Rourke (top); Krysta Doefler, courtesy *Fine Homebuilding* magazine, ©The Taunton Press, Inc. (bottom left, bottom center, and bottom right)

pp. 106–107: courtesy Mannington Mills, Inc., www.mannington.com (top and bottom)

p. 109: Eric Roth, design: JW Construction, Lexington, MA (top); Brian Vanden Brink, design: Polhemus Savery DaSilva Architects Builders (bottom)

p. 110: Brian Vanden Brink, design: Hutker Architects (top); ©CarolynBates.com, interior design: Lisa Proli and Josh Rourke; contractors: Linda and Michael LaCroix, Aspen Construction Services Corp. (bottom)

p. 111: Randy O'Rourke (top left); Susan Teare, builder: Silver Maple Construction (top right); Brian Vanden Brink, design: Houses & Barns by John Libby/Morningstar (bottom left); Andrea Rugg Photography/Collinstock, design: Jen Seeger Design, Rosemary Merrill Design (bottom center); Brian Vanden Brink, design: Carl Solander, Architect (bottom right)

p. 112: Eric Roth, design: Margo Evans, Boston (left); Brian Vanden Brink, design: Polhemus Savery DaSilva Architects Builders (right)

p. 113: Eric Roth, design: Warner Cunningham Architecture, Brookline, MA (top left); Hulya Kobalas, design: CWB Architects (top right); Eric Roth, Lisa Tweed Design, Boston (bottom left); Eric Roth, design: Trikeenan Tileworks, Swanzee, NH (bottom right)

p. 114: ©CarolynBates.com, architect and interior design: Nils Luderowsk (top); Jo-Ann Richards, Ines Hanl, The Sky Is The Limit Design (bottom)

p. 115: Brian Vanden Brink, design: Dominic Mercdante Architect (top); Brian Vanden Brink, design: Polhemus Savery DaSilva Architects Builders (bottom)

p. 116: Randy O'Rourke (top and bottom)

p. 117 Randy O'Rourke (top and bottom)

CHAPTER 5

p. 118: Brian Vanden Brink, design: Morningstar Marble & Granite

p. 120: Chipper Hatter, design: Home Improvements Group, Chris Dreith—designer (top); Andrea Rugg Photography/Collinstock, design: Anchor Builders (bottom)

p. 121: Brian Vanden Brink, design: Hutker Architects (top); Eric Roth, design: Stern McCafferty Architects, Boston (bottom left); Brian Vanden Brink, design: Seimasko + Verbridge (bottom right)

p. 122: Brian Vanden Brink

p. 123: Chipper Hatter, design: Shiny Bones, Myca Loar—designer

p. 124: Brian Vanden Brink, design: Hutker Architects (top); Randy O'Rourke (bottom)

p. 125: Chipper Hatter, design: CM Natural Design, Corine Maggio—designer

p. 126: Brian Vanden Brink, design: Maine Kitchen Designs

p. 127: Eric Roth, design: Butz+Klug Architects, Boston (top); Susan Teare, construction: Conner & Buck Builders, design: Annette Tatum (bottom)

p. 128: Virginia Hamrick Photography/ Collinstock, design: Candace M. P. Smith Architect

p. 129: Brian Vanden Brink (top left); ©CarolynBates.com, interior design: Donna Sheppard Interiors, contractor: Tom Sheppard (top right); Eric Roth (bottom)

p. 130: Eric Roth, design: Jonathan Poore Architects, Gloucester, MA

p. 131: Randy O'Rourke (top); Undine Prohl, design: Alter Studio (bottom left); Eric Roth, design: ZEN Associates, Boston (bottom right)

p. 132: Randy O'Rourke

p. 133: Susan Teare, design: Hart Associates Architects, Kate Daly project manager, Gilman Guidelli and Bellow General Contractors. Color-changing temperature-sensitive floor tiles from http://www.dudeiwantthat.com/

p. 134: Brian Vanden Brink, design: Hutker Architects

p. 135: Randy O'Rourke (left and right)

p. 136: ©CarolynBates.com, interior design and contractor: Sandy Lawton, ArroDesign (left); Randy O'Rourke (right)

p. 137: Brian Vanden Brink, design: South Mountain Co. (left); Brian Vanden Brink, design: Breese Architects (right)

p. 138: Randy O'Rourke

p. 139: Randy O'Rourke (top left, top right, bottom)

CHAPTER 6

p. 140: Randy O'Rourke

p. 142: ©CarolynBates.com, architect: Elizabeth Brody

p. 143: Brian Vanden Brink, design: Polhemus Savery DaSilva Architects Builders (top); Eric Roth, EJ Jaxtimer Builders, Osterville, MA (bottom)

p. 145: courtesy Broan Nu-tone (top); Randy O'Rourke (bottom)

p. 146: courtesy Broan Nu-tone (top left); courtesy Broan Nu-tone (top right); Krysta Doefler, courtesy *Fine Homebuilding* magazine, ©The Taunton Press, Inc. (center and bottom)

p. 147: courtesy Broan Nu-tone (top); Undine Prohl, design: Safdie Rabines Architects (bottom left); Krysta Doefler, courtesy *Fine Homebuilding* magazine, ©The Taunton Press, Inc. (bottom right)

p. 148: Rob Karosis Photography/ Collinstock, design: Whitten Architects (top); Krysta Doefler, courtesy *Fine Homebuilding* magazine, ©The Taunton Press, Inc. (bottom left and bottom right)

p. 149: Randy O'Rourke

p. 150: Susan Teare, design: Peregrine Design/Build

p. 151: Eric Roth, design: Butz+Klug Architects, Boston

p. 152: Randy O'Rourke

p. 153: Randy O'Rourke (top, bottom left, bottom right)

CHAPTER 7

p. 154: Emily Followill Photography/ Collinstock, design: Elizabeth Spangler Design

p. 156: Eric Roth (top); Brian Vanden Brink, design: Rob Whitten Architects (bottom)

p. 157: photography by Troy Theis, design: Martha O'Hara Interiors

p. 158: Eric Roth, design: Hutker Architects, Falmouth, MA, Martha's Vineyard Interior Design, Vineyard Haven, MA

p. 159 Eric Roth, design: Hutker Architects, Falmouth, MA (top); Chipper Hatter, design: Kitchen and Bath Concepts, Sara & Jay Meloy—designers (bottom left); Randy O'Rourke, design: Hudson Valley Preservation Corp. (bottom right)

p. 160: Hulya Kolabas, decor by Guillaume Gentet Inc.

p. 161: Brian Vanden Brink, design: Drysdale Design Associate (top left); Brian Vanden Brink, design: Hutker Architects-Aronson Residence (top right); Andrea Rugg Photography for U + B Architecture & Design (bottom)

pp. 162–163: Brian Vanden Brink, design: Dominic Mercdante Architect-Dyke

p. 162: Brian Vanden Brink (bottom)

p. 163: Susan Gilmore photographer (left); Greg Premru Photography (right)

p. 164: Andrea Rugg Photography/ Collinstock, design: CF Design Ltd.

p. 165: Eric Roth, design: Martha's Vineyard Interior Design, Vineyard Haven, MA (left); Ryann Ford Photography, design: CG& S Design Build (right)

p. 166: Rob Karosis Photography/ Collinstock, design: C. Randolph Trainor Design (left); Brian Vanden Brink, design: Hutker Architects (right)

p. 167: Eric Roth, design: Butz+Klug Architects, Boston (top left); ©CarolynBates.com, architect and interior design: Nils Luderowski (top right); courtesy Rev-a-Shelf LLC (bottom)

p. 168: Eric Roth, design: LDa Architects, Boston

p. 169: Chipper Hatter, design: Kitchen and Bath Concepts, Sara & Jay Meloy—designers (top left); Mark Lohman, design: Caroline Burke Designs (top right); photography by Troy Theis, design: Meghan Kell Cornell, AIA, and Lucy Interior Design (bottom)

pp. 170–171: Randy O'Rourke

p. 170: Randy O'Rourke (bottom)

p. 171 Randy O'Rourke (right)

p. 172: Brian Vanden Brink, design: Jackie Silverio Architect

p. 173: Eric Roth, design: Hutker Architects, Falmouth, MA (top); Eric Roth, design: Robert Bowman Builders, Falmouth, MA (bottom left); Randy O'Rourke (bottom right)

p. 174: Susan Teare, builder: Stewart Construction, architect: Keefe and Wesner Architects (top); Susan Teare, design: Hart Associates Architects, builder: Kistler and Knapp Builders (bottom)

p. 175: Randy O'Rourke (top left); Brian Vanden Brink, design: Hutker Architects (bottom left); Randy O'Rourke (right)

p. 176: Susan Teare, builder: Conner & Buck Builders, design: Pill-Maharam Architects (top); Susan Teare, general contractor: Red House Building, interior design: Cushman Design Group (bottom)

p. 177: Hulya Kolabas, design: Sam Allen Interiors (top left); Chipper Hatter, design: KW Designs, Kristianne Watts—designer (top right); Emily Followill Photography/Collinstock, design: Beth Kooby Design (bottom left); Rob Karosis Photography/ Collinstock, design: Smith & Vansant Architects, DPF Interior Design (bottom right)

p.178: Randy O'Rourke (top and bottom)

p. 179: Randy O'Rourke (top and bottom)

FLOORS,

Choosing the right surface materials for your bathroom will define the room's style

WALLS, AND

while creating a space that's durable, comfortable to use, and easy to clean.

CEILINGS

Nothing makes a bigger impact on the look of your bathroom than your choice of materials for the floor, walls, and ceiling surfaces. Naturally, you want to choose surface materials that fit your tastes and the style of your home. However, it's equally important to consider how the bathroom will be used when making these decisions. Some materials will withstand heavy use better than others. Even more crucial is being aware of how much a surface will be exposed to moisture.

Tile is the first material that comes to mind when we think of bathrooms. It's beautiful and durable, and it comes in many shapes, colors, and textures. Tile can also get expensive, but you needn't cover the whole bathroom with it. You can use tile judiciously to cover areas that get wet. Reasonably priced tile in the shower and perhaps as a sink backsplash combined with walls coated with good-quality paint can result in a gorgeous bathroom that won't break the budget.

Some surface materials, most notably wood and wallpaper, are more prone to water damage than others. For example, a powder room with just a sink and toilet doesn't get much water on the floor, so a wooden floor wouldn't be a problem there. Taking a bath doesn't produce much steam, so wooden walls might be fine in a room with a tub but no shower. However, the floor will get wet when people exit a tub, so you might choose a more impervious material than wood for the floor, especially if the tub will be used to bathe kids. In most cases, you wouldn't use wood inside a shower. If the whole bathroom is likely to fill with steam, be aware that some species of wood—especially hardwoods— are more likely than others to buckle if exposed to excess moisture from shower steam.

Of course, in addition to thinking about the look you want to achieve and materials that are appropriate for the situation, you'll be thinking about your budget. Surface materials vary widely in cost, but whatever your budget, you'll find materials that will look great and perform well in your bathroom design.

above • Great bathroom design can successfully blend the old with the new. Here, an old-fashioned bathtub is surrounded by small octagonal tiles set in an eclectic free-form design. Paint picks up the motif as the walls move away from the wet area, while the floor of small blue octagonal tiles pulls it all together.

facing page top left • The warmth of wood framed by large white moldings, a floor of white octagonal tiles with dark grout, and traditional fixtures give this handsome bathroom a classic look. To add visual interest, the wood wall stops well short of the ceiling and becomes a partition wall for the shower.

facing page top right • Even with large tiles you don't need to color within the lines. Tiles of various sizes and colors are skillfully cut to turn this bathroom into a work of art.

facing page bottom • This bathroom is a study in how a variety of surface colors and textures can work in harmony. A pathway of pebbles set in mortar meanders through a stone tile floor to cover the front of the tub platform. Copper-colored walls lead to a rustic exposed-beam ceiling. The step to the tub and the top of the platform feature the same tiles as the floor, helping to tie the composition together.

Floors

For floors that will get wet on a regular basis, you want to use materials that are impervious to water. For that you can't beat tile, whether porcelain, ceramic, or glass. Concrete also works well if properly sealed. Vinyl flooring, while less durable, is quite inexpensive and resists water—especially in sheet form, which has no seams for water to seep into.

If the floor won't be subjected to lots of water, just about any flooring material is worth considering, including solid wood, engineered wood, linoleum, cork, or laminate. If your bathroom is large, one strategy is to tile the shower area or under the tub and then use another material for the floor areas that will be subjected to less water. The only conventional floor material that should absolutely be avoided in the bathroom is carpet, with the exception of bath mats or small throw rugs that can easily be lifted for cleaning and drying.

above • Sealed concrete makes a great bathroom floor. It can be tinted in any color, or it can be left its natural grey to let other design elements take center stage.

facing page top • Grey floor tiles with pebble inserts work with a lighter grey wall with painted pebbles to make the warm wood details pop. A glass shower panel looks like a wall of twigs.

facing page bottom • A field of tiles that appear woven together is a classic look that works perfectly with an old-fashioned freestanding tub.

TILE

It's no wonder tile is the most popular material for bathroom surfaces, especially walls and floors. Its broad range of colors, sizes, and patterns, along with its water and stain resistance and durability all help make it an excellent choice.

Tile can fit into any design theme. For a traditional look, you can use white tiles with smaller black accent tiles. Another option is to create a subdued field of neutral-color tiles that will let other features in the bathroom pop. Using tiles of the same color but different sizes can create a subtle transition between the floor inside and outside a curbless shower or between walls and floor.

Or you can use boldly colored tiles as a focal point. Introducing a few fancy tiles into an otherwise plain floor adds lots of visual interest without adding a lot of cost. A pattern of mosaic tiles can be bordered with larger tiles of one color to look like a rug. Really, when it comes to tile, your own imagination and taste—and of course your budget—are the only limits.

Most of the tile available is porcelain or ceramic. Porcelain is actually a type of ceramic that's a bit denser and has color throughout rather just on the surface. As a result, porcelain tile is a bit more durable and chipping damage would be less obvious because you won't see a different color. Both porcelain and ceramic are highly resistant to water absorption and stains.

Should a tile chip or crack, which can happen even though tile is very durable, it can be replaced without disturbing the rest of the floor. This is a real advantage, but only if a tile of the same color is available. Manufacturers often change colors and patterns, so it's a good idea to stash a few spares away when the tile is originally set.

above • Each small, square black tile is surrounded by four white tiles of the same size to create the classic look that works perfectly with the white tile wainscot and the old-style freestanding tub.

facing page • You needn't spend a fortune on tiles to create visual interest. Here, relatively inexpensive tiles of different style and color are used on the wall and floor while more expensive tiles are reserved for a floor border near the tub.

Glass floor tiles are available and are quite beautiful as well as resistant to water and stains. However, they tend to get scratched over time, which degrades their appearance. Concrete tiles are durable, and they can be created in any shape, size, or texture or, if you like, your floor can be one concrete slab. Concrete is porous, however, and must be sealed—not against moisture, which won't hurt it, but against stains and ground-in dirt. Terra-cotta tile is also porous and should be sealed.

Whatever tiles you use, the grout joints will be porous and will need to be sealed against water, stains, dirt, and the formation of mildew. Any sealant you use—on tiles as well as grout joints—will eventually wear off and need to be reapplied.

One thing tile doesn't like is too much flex in the floor. The larger the tile, the greater the risk of cracking. Before deciding to install tile, discuss with your contractor whether your floor is stiff enough. Some floors, especially in an older home, may need to be reinforced. In some cases, installing a special membrane, such as Schluter-Ditra, between the tile and subfloor may do the job by isolating the tile from substrate movement.

The biggest drawback to tile is that it can be slippery when wet. If that is a concern, look for a tile with a dynamic coefficient of friction (DCOF) of at least 0.42. The higher the number, the more resistant the surface is to slips. Tiles that are textured or unglazed usually have a higher DCOF. Shiny glazed tiles tend to be more slippery.

Tiles with a raised pattern add a basketweave texture to this floor.

Keeping Tile Warm

One drawback to tile in the bathroom is that it can be cold under bare feet in the winter months. If your home already has hot-water radiant heat in the floors, you won't have this problem. If conventional hot-water radiators heat your home, it may be possible to tap into that system to install hot-water radiant heat in the bathroom floor.

An alternative for homes with heating systems that don't use hot water is to install an electric heating mat under the tile. These mats contain a grid of electrical wires that are controlled by a thermostat or timer. Mats, such as those made by Nuheat™, are bedded in a layer of thinset adhesive with the tile installed on top of that. They are relatively inexpensive to run, especially when used for short periods of time after a bath or shower.

TIP: If you are installing hot-water radiant heat in your bathroom, it's relatively easy to continue the hot water loop into a rack that will keep your towels toasty. Ask your contractor about this option.

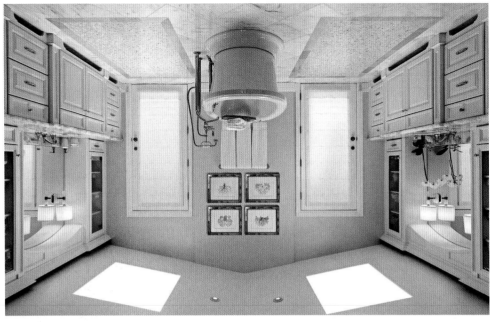

above • **Large dark tiles emphasized by light grout give this floor a bold look. For more visual interest, the tiles are set in a running bond—the end joints in each course fall in the middle of the courses next to it.**

right • **Fields of patterned tiles surrounded by blue borders set into a larger field of light-toned marble tiles evoke carpets and help to define the spaces. This design also leads the way to the separate showers in this his-and-hers master bath.**

STONE

The special beauty of stone comes from the fact that no two pieces are alike. Within the same type of stone, sometimes even in a single stone tile, you'll find an exciting variation due to natural imperfections, veining, even fossils. Stone is generally more expensive than other tile, but it offers a rich and lively appearance that no man-made material can match. For a completely different look, you can use stone in the form of smooth pebbles set into mortar.

Stone is available in a broad range of colors, from the light hues of limestone to the dark gray of blue slate. Some types such as granite are available in different colors. You'll have no problem finding stone that looks great with other colors in the room.

Textures vary too, from the elegance of a light-colored polished marble to the rustic look of travertine that has been tumbled to soften its edges. As with tile, stones can be mixed and matched with one type, color, and/or size on the floor and another in a shower or with two types of stone laid in a pattern on the floor.

Like man-made tile, stone is durable, wear-resistant, and easy to clean. However, all stone is porous to some degree and should be sealed. The more porous the stone, the softer it is and the more sealant it will require. Harder, less porous stone such as granite and some types of slate will absorb less sealant and will stand up better to heavy traffic than softer stones such as marble, travertine, or limestone.

If you are interested in building green, look for stone that is quarried nearby to minimize the costs and environmental impact of shipping.

Stone is trickier to lay than man-made tile and should be installed by a pro. And stone is even less tolerant of flex in the subfloor than tile is, especially if the stone pieces are large. Thick pieces of stone set in a mortar bed may need extra floor framing, such as joists set 12 in. on center instead of the more typical 16 in. on center. Be sure to discuss this well in advance with your builder.

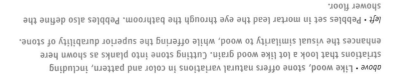

above • Like wood, stone offers natural variations in color and pattern, including striations that look a lot like wood grain. Cutting stone into planks as shown here enhances the visual similarity to wood, while offering the superior durability of stone.

left • Pebbles set in mortar lead the eye through the bathroom. Pebbles also define the shower floor.

Facing page top • Several types of stone are used to great effect in this bathroom, including a green marble floor and counter. The shower enclosure has light-colored stone tiles in three sizes complemented by bands of narrow stone tiles above the seat and window.

Facing page bottom • A polished marble floor is the height of elegance. It can also be quite slippery. This one features a graceful border and flows right into the curbless shower.

SOLID WOOD AND ENGINEERED WOOD

Solid wood flooring, as the name indicates, is milled from wood planks. Engineered wood flooring consists of high-quality plywood with a layer of solid wood glued on top. Whether solid or engineered, wood offers a warm, natural beauty but isn't the most practical choice in many bathroom situations.

In general, solid wood is not a great choice for a floor that will be exposed to a lot of fluctuation in humidity. You can think of a piece of solid wood as a bundle of straws glued together—the straws are the grain. When exposed to humidity, whether from a sticky summer day or the steam of a hot shower, the straws absorb moisture from the air, causing planks to expand across the grain. When the air is dry, the straws release moisture and the boards contract. This expansion and contraction causes the gaps between the boards to grow and shrink. In cases of extreme humidity, boards can expand enough to cause the floor to buckle.

Engineered wood is more dimensionally stable than solid wood. That's because its base is made of several plies of wood glued together with alternating grain direction, eliminating the tendency to expand and contract.

Neither solid nor engineered wood is a good choice for a floor that will get wet from people stepping in and out of a shower or bath or from condensation dripping off a toilet tank on a warm day. Unless the floor is well sealed, the wood will absorb water and eventually rot. Today most solid and engineered floors come prefinished. Still, it is difficult to completely seal a wood floor because the joints expand and contract.

Wood is fine to use in a powder room. And wood might be practical enough under a claw-foot bathtub, especially if there is no shower, because there is no tub-to-floor joint to collect water. If you have your heart set on a wood floor, consider using tile around the shower and/or tub with wood for the rest of the floor.

A good-quality floor paint increases the durability and water resistance of a wood floor, and painting an existing floor costs a lot less than installing tile. The stenciled motif on this floor adds lots of visual interest.

You can't beat the look of a wood floor in an old house. Combining practicality with beauty, tile is used below and behind the tub and at the shower door.

The warm orange beauty of this pine plank floor pops out against the muted tones of the other surfaces in this bathroom.

Although technically a grass, bamboo is considered another form of engineered wood flooring. Bamboo strips are woven together and then fused under pressure. The result is a tough surface that many people find quite attractive. Like other engineered wood flooring, bamboo is dimensionally stable and has the same problems with moisture. Bamboo is considered a renewable resource, something to consider if you are building green.

On the plus side, wood is warmer and more comfortable underfoot than tile or stone. It is also much more forgiving of flex in the floor than tile or stone, and in fact will help stiffen a bouncy floor. Solid wood can be sanded and refinished numerous times. Engineered wood, depending on the thickness of the top layer, can be only be sanded one or two times.

The vast majority of wood floors are made of hardwoods; solid flooring is available in many species, offering more options than engineered flooring. If you are looking for a really rustic or old-fashioned appearance, you can use pine planks. As a softwood, pine is more susceptible to dents than hardwood—perhaps not a big issue for a bath where the family will mostly be padding around barefoot or in slippers. Also, if you have been admiring old honey-hued pine floors, be aware that old growth pine is no longer available unless you salvage antique boards. New pine lacks the color and distinctive grain of old pine, and it is more susceptible to rot. Pine, along with other softwoods such as cypress and fir, is considered a renewable resource.

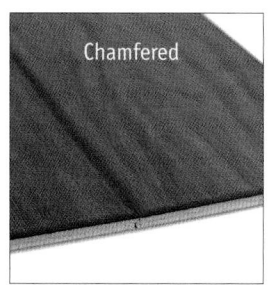

Square-edged

Chamfered

Beveled

Laminate flooring can be made to look like natural wood but can be damaged if the inner core gets wet.

VINYL

From a purely practical point of view, vinyl is an excellent choice for a bathroom floor. It is impervious to water when properly installed. It's also simple to install, and while it's not as durable as tile or stone, it's significantly less expensive and it feels softer and warmer underfoot.

Vinyl is available in sheets, which is the least expensive and easiest form to install. It is also available as tiles and, most recently, in planks designed to mimic wood. Vinyl made to look like ceramic tile, stone, and wood has gotten more realistic in recent years, but it still doesn't look or feel as natural as the real thing.

The better grades of vinyl are inlaid—the color and pattern go all the way through the flooring, so even in high-traffic areas, wear won't be as obvious.

Be aware that some vinyl flooring contains phthalates, chemicals used to make the vinyl more pliable. Phthalates are suspected of causing cancer, birth defects, and developmental disorders. There is no conclusive evidence indicating enough phthalates can leach from vinyl flooring to cause any harm, yet several major retailers including Home Depot®, Lowes®, Menards®, and Lumber Liquidators® have stopped stocking vinyl flooring that contains these chemicals. You might want to look for phthalates-free vinyl flooring, especially if you have small children.

above • The sheet vinyl on this floor is made to look like small tiles. It's impervious to water and soft underfoot. It's less durable than real tile but also much less expensive and easier to install.

right • Vinyl tiles are a bit more work to install than sheet vinyl but look more like ceramic tiles.

LAMINATE

Laminate flooring consists of a tough layer of plastic over a man-made substrate, such as fiberboard. Between the plastic and the substrate is a printed film that gives the flooring the look of another material, such as tile, wood, or stone. To enhance the realism, some laminate flooring is embossed to emulate wood grain or the grout in tile.

Most laminate flooring is literally a snap to install. The floor "floats," meaning it's not glued or nailed down, and has a tongue-and-groove system that snaps together without glue.

The substrate of laminate flooring will swell if it gets wet, so you might not choose laminate for a kid's bath or other situation where the floor will be subjected to a lot of water. It's also important to carefully caulk around the perimeter of the floor to keep water from seeping in. Be aware that some manufacturers void the warranty if laminate is used in a bathroom. One exception is SpillShield™ from Mannington®, a factory-applied coating on that company's laminate flooring that is specifically designed for use in areas that get wet.

LINOLEUM AND CORK

Vinyl flooring nearly killed the linoleum industry, but an increasing interest in eco-friendly building materials has revived interest. Linoleum is made of linseed oil and other natural materials, such as pine rosin and ground cork and wood, usually with a burlap backing. Like vinyl, linoleum is resilient underfoot. It's available in sheets, tiles, and planks and in a variety of colors and patterns.

Linoleum is water resistant but it's not impervious to water, and water penetration can occur. If linoleum is used in the bathroom, the seams should be heat-sealed against moisture. Some manufacturers don't recommend using their product in wet situations; in some cases, using linoleum in a bathroom may void the warranty.

Cork is another natural material that comes in sheets, tiles, and planks. It is the bark of the cork oak tree, which is harvested periodically without harming the tree. It contains tiny air bubbles that make it resilient underfoot and also make it a good thermal and sound insulator. It's naturally light in color and can be stained to darker hues. Like linoleum, cork is not impervious to water and must be carefully installed and sealed if used in the bathroom.

Flooring

Water resistance is key, especially in a bathroom where splashing in the tub or around the sink will occur, such as a kid's bath. If the floor won't be subjected to much water, then color, texture, and feel underfoot may be more important considerations. In all categories of flooring, prices vary widely. Basic ceramic tile available at home centers and tile stores, for example, costs little more than vinyl tile, while some glass and art tile can get very pricey.

VINYL
$

- Resilient and comfortable underfoot.
- Great variety of patterns and colors.
- Available in sheets, tiles, and planks.
- Inexpensive but not as durable as stone or ceramic tile.
- Highly resistant to stains and impervious to water when properly installed.
- Vinyl tile can be installed by do-it-yourselfers, but where standing water is likely, it's best to minimize joints by using sheet vinyl.

LAMINATE
$

- Snap-together installation is fast and easy.
- Inexpensive.
- Sold in many colors and patterns to look like tile, stone, or wood.
- Top wear layer is durable, but fiberboard core will swell if exposed to water. Keep all perimeter edges sealed and caulked.
- Standing water needs to be cleaned up promptly.
- Some manufacturers void warranty if laminate is used in a bathroom.

SOLID WOOD
$-$$

- Warm underfoot.
- Comes in many species; can be stained any color.
- Visually appealing and well suited to period homes.
- Not a good choice around tubs, sinks, and toilets unless the finish is maintained and edges are caulked and sealed.
- Shrinks and expands with changes in indoor humidity. Gaps between boards can widen in winter. In warm, humid weather, boards may cup and the floor may even buckle if the air was dry when the boards were installed.

TILE
$-$$

- Huge variety in color, texture, and surface appearance.
- Offers great design flexibility.
- Extremely durable.
- Requires very little maintenance.
- Ceramic and porcelain tiles are resistant to stains and cracking.
- Requires stiff floor framing to prevent cracking.
- Without radiant-floor heating is cold underfoot.
- Grout should be sealed to prevent stains and mildew.

ENGINEERED WOOD
$$

- Made from a layer of wood over a plywood core.
- Gives the appearance of solid wood but will shrink and expand less with changes in humidity.
- Like solid wood, not the best choice in a bathroom if frequent wetting and standing water are likely.

LINOLEUM AND CORK
$$

- Resilient flooring that is more environmentally friendly than vinyl.
- Warm and cushiony underfoot.
- Made from natural materials.
- Naturally biodegradable.
- Available in tile or sheet form.
- More expensive than vinyl and requires more care.

STONE
$$-$$$

- Long-wearing natural material with great visual appeal.
- Wide range of colors on the market.
- Some polished stone is very slippery when wet.
- Thick stone may require heavier-than-usual floor framing.
- Most types of stone should be sealed periodically to prevent stains.
- Grout between stones may stain if left unsealed.

When you want natural light from outside or an adjoining inside space but you also need privacy, a wall of glass block can be the solution.

Walls and Ceilings

As with other bathroom surfaces, it's important to think about how the bathroom will be used when choosing a wall or ceiling surface. If it's a guest powder room or even a master bath with no shower to generate steam, feel free to use any finish you would use in the rest of the house, including wallpaper.

Shower walls, of course, should be covered with an impervious surface, such as ceramic or porcelain tile, that is durable, easy to clean, and impenetrable to water. Though more expensive, glass tile is ideal for walls, where it won't be subjected to the scratching and wear that happens on floors. Properly sealed, stone is another fine choice.

Outside the shower, wall finishes that will be subjected to steam and splashes should be water resistant; because water won't stand on walls and ceilings, the finish need not be completely impervious.

The simplest choices are painted moisture-resistant drywall or painted plaster, if that's what is already in place. Here, the choice of paint is important. The glossier the paint, the more it will prevent water from soaking in. And glossier paints are easier to wipe clean. For this reason, semi-gloss and high gloss paint have traditionally been used in bathrooms. Today, if you prefer the look of matte walls, you can buy matte paint that is specially formulated to resist moisture. Look for cans labeled "bath" or "bath and spa."

Wood is another acceptable choice for bathroom walls and ceilings. You can paint wood as you would drywall or plaster, or you can apply a water-resistant clear finish, such as polyurethane.

left · White walls and pale blue tile create a soothing mood and draw the eye to the view through the Palladian windows next to the tub.

REPAIR VS. REPLACE

When renovating a bathroom, you might have to decide whether to repair or replace wall and ceiling surfaces. Demolition makes a mess, not only in the bathroom but also, potentially, in the rest of the house. Plus, it adds to the length and cost of the job.

One reason to leave existing walls in place is to preserve valuable architectural details or rare tiles. Short of that, though, it usually makes sense to strip the walls to the framing. In older homes, the insulation in exterior walls is often inadequate. Moisture migrating into wall cavities over the years may have promoted mold or even resulted in structural decay of framing or sheathing. Stripping the walls provides the opportunity to inspect the structure, not only for damage but also to make sure the floor framing can carry the load of tile or perhaps the bigger tub you are planning. In addition, the plumber can get a good look at the pipes to make sure nothing needs replacing.

Tearing out also greatly simplifies wiring upgrades—adding receptacles, for example, or replacing old cable that lacks a ground wire. Removing the ceiling makes it easy to enhance lighting and to add a fan that's vented to the outside if the bathroom doesn't already have one.

In a gently used powder room you can use the same materials you would use anywhere in the house. Here, a simple wooden table supports a vessel sink, blending beautifully with the wood floor and wall trim. White paint warmed to a linen color looks great with natural wood trim.

right • The warm tones of the tiles used in the wet areas of this bathroom blend beautifully with the reclaimed planks behind the sink. A stripe of smaller tile adds a focal point that leads the eye through the space.

Large slabs of stone hold their own with the massive exposed timber frame of this house. A stained glass panel provides privacy but also can be slid aside to enjoy the view or to clean the window.

White subway tiles on the walls (and ceiling) and tiny octagonal tiles in many hues of blue on the floor create a classic look in this shower. Using the same colored tiles in the shower niche breaks up the white field.

Green-tinted glass tiles define this shower space. Glass tiles are an excellent choice for wet areas because they are impervious to water and easy to clean. They can scratch and scuff under foot traffic, but in a shower stall, where feet are always bare, they'll hold up fine.

Appropriate use of materials adds texture to the subdued, timeless look of this bathroom—tile on the floor and in the shower, a marble countertop, and wooden wainscot on the walls. A subtle strip of lighter blue creates a transition between the white wainscot and the blue above.

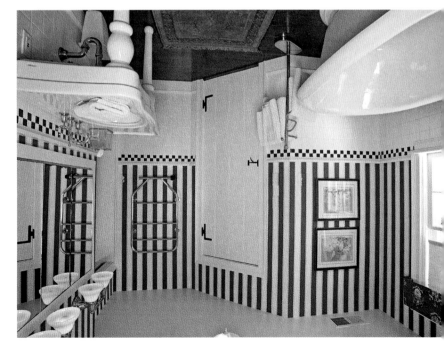

Wallpaper should be used judiciously in the bathroom. Here, a tile wainscot covers areas that might get splashed around the tub and sink. Besides being practical, the effect is much more interesting than tile or wallpaper alone. With no shower steam, the wallpaper will hold up just fine.

Small and narrow iridescent glass tiles create a striking wainscot behind this sink.

A cost-savvy custom look is achieved by accenting plain white colored tiles set at a 45° angle in this shower enclosure.

TILE

Virtually endless design possibilities coupled with durability and imperviousness to water make tile a top choice for bathrooms. You can combine various shapes, colors, and sizes to perfectly suit your taste and the style of your home. Tile can get pricey, especially in stone, but it doesn't have to be. Inexpensive tiles from your local home center or tile store are just as durable as top-end tiles. One way to achieve a custom look is to strategically place a few handmade or brightly colored tiles on a plain field of inexpensive tiles.

Another way to save money while also adding visual interest is to use tiles just partway up the wall in what's known as a wainscot. This technique originally was

executed in wood in dining rooms to protect walls from chairs when folks slide back from the table. Likewise, you can use a tile wainscot to protect the part of the wall that will get splashed. Of course, you can use a wood wainscot in the bathroom, too.

In general, wainscot looks best when it covers more than the bottom half of the walls. In determining the height of wainscot, tile experts Lane and Tom Meehan, authors of *Working with Tile* (The Taunton Press, 2011), suggest the top of the wainscot be aligned with the top of the vanity backsplash. Wood wainscot is usually topped with chair rail, a molding that extends farther from the wall than the wainscot. Chair rail also is available as tile.

Tiles in multiple sizes and muted, almost metallic tones set with random joints add wonderful texture.

A fancy tile wainscot surrounds a claw-foot bathtub to create an old-timey elegance. The wainscot features a row of green pencil tiles, topped by a course of tiles with a leaf motif, and finished off with chair-rail tiles. The blue walls and white ceiling visually tie in to the tub.

A wainscot of small square blue tiles with contrasting grout and no chair rail creates a striking modern look.

A decorative paint finish complements the square tiles in hues of blue, green, and purple used in areas that will get wet. The tiles used as a baseboard look great and are also a practical choice. The joint between tile and a wooden baseboard can be vulnerable to rot if not carefully sealed. Tile is also easier to clean.

WOOD

Whether painted or coated with a clear finish to celebrate its natural beauty, wood is a great way to add texture and color to your bathroom design. From an elegant frame-and-panel wainscot to rustic reclaimed barn boards, wood can be incorporated into any design.

Perhaps the most common use of wood in the bathroom is a wainscot of beaded tongue-and-groove boards. Of course, you can run the boards all the way up to the ceiling. And covering just the ceiling with beaded boards can help tie a room together. Most often the boards are run vertically on the walls, but installing the boards horizontally might fit perfectly into your design. You can save money and time by using plywood panels incised to look like beaded board, but the detail is shallower than the real thing.

Generally, wood walls and ceiling invoke a traditional look, but the material can be incorporated into very modern designs as well. Using plywood panels instead of planks offers the opportunity to create seamless fields. Incorporating panels veneered with a species such as birch or maple that has subtle grain patterns can provide a sleek and modern look. If you want to go green, take a look at products such as Echo Wood®, which has surface veneers made of reconstituted wood pulp.

Because wood species vary in color and grain patterns, what you choose will be an important part of your design scheme if you'll be using a clear finish. Douglas fir, for example, has orange tones with prominent grain, while cherry has very subtle grain and a reddish color that darkens with age. Walnut is brown with striking grain, while birch and maple are light in color with subtle grain.

If you're going to paint wood, choose clear grades of lumber. Sealing primers that contain shellac are designed to seal in knots, but the resins have a way of leaching through.

top right · Horizontal beaded boards with a natural finish continue up the ceiling to evoke a modern, nature-inspired feel. Simple white subway tiles used around the tub are practical and provide a visual counterpoint to the wood.

right · Plywood panels provide a sleek and seamless way to cover walls and ceilings with wood. Another, more ecologically friendly alternative is to use panels made entirely of reconstituted wood fiber, such as those shown here.

Walls and Ceilings

DRYWALL
$

- The least expensive wall surface.
- Installation can be handled by a do-it-yourselfer.
- Can be repainted, making it easy to change the appearance of the room.
- Not a very robust surface but can be repaired easily.
- Use water-resistant board to minimize the risk of water damage from moisture.
- Not appropriate for use in showers or tub surrounds.

PLASTER
$$

- A more expensive option than drywall, and usually reserved for historic restorations.
- Skilled labor to apply may be hard to find locally.
- Makes a hard, durable finish.
- Surface can range from textured to glassy smooth.

TILE AND STONE
$-$$$

- Very durable.
- Wide variety of colors and textures.
- Impervious to water, so can be used in showers and on tub surrounds.
- Grout should be sealed.
- Nonporous tile surfaces are easy to keep clean.

WOOD AND WOOD LOOK-ALIKES
$-$$

- Good durability as long as it is painted or sealed and the surface is maintained.
- Fairly easy to repair if damaged and easy to refinish.
- Can be installed as tongue-and-groove planks or flat panels.
- May be damaged in high-moisture areas. Not a good choice for a shower or tub surround.

top left • There is nothing more classic looking than a beaded-board wainscot painted white.

left • A coffered ceiling and walls covered with beaded boards and a wooden floor made to look like a boat deck give this bathroom a nautical look. The porthole mirror and sailboat painting continue the theme.

An Un-Bathroom Vibe

Rather than our usual notion of a bathroom, this space feels more like a room that happens to include a bath and a shower. Several design touches contribute to the decidedly un-bathroom-like vibe of this master bath. For starters, there's no vanity or, in fact, any built-in cabinetry at all. The sink vanities, along with a makeup station, are in a separate space just off the bathroom. The toilet has its own room with a pocket door.

Other touches include a chandelier you'd be more likely to find over a dining table than over a bathtub and a freestanding display cabinet that you'd expect to find in a living room.

Nothing says bathroom more than the use of wall tile—that makes sense since tile is the most practical surface for spaces that get wet. And tile *is* used in this bathroom, but it's reserved for the walls of the large walk-in shower and the floors inside and outside the shower. In front of the tub the field of travertine marble floor tiles is inset with tiles that evoke the look of an area rug you might find in a living room.

The windows in the shower existed before architect Paul Hannan of SALA Architects redesigned this space, but they were in a closet. Turning the closet into a shower with a glass partition transformed a dark space into a spacious bathing room that's flooded with light. This makes lingering in the soaking tub an even more pleasurable experience.

Two vanities flank a makeup area in a space between the master bedroom and master bath. The sliding doors at right lead to the master closet.

The marble "area rug" is set in a field of tiles that also are marble.

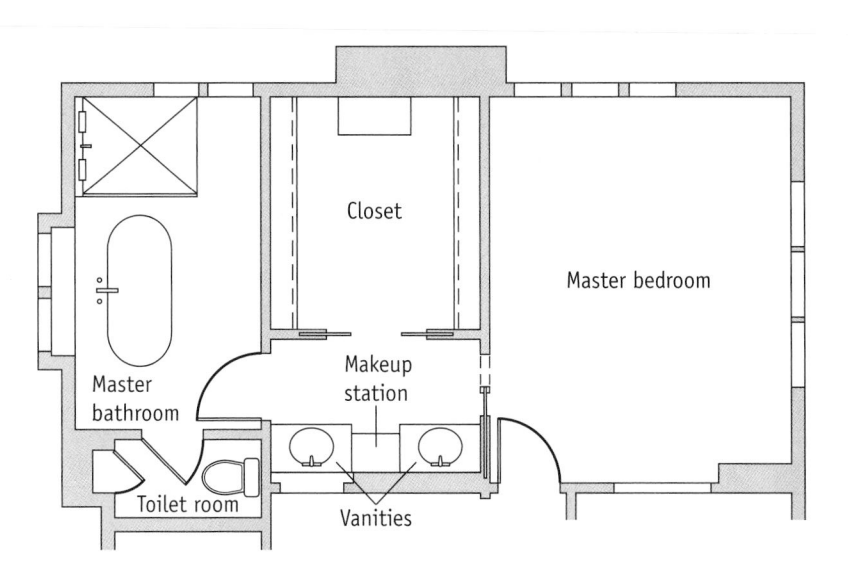

This master bath has a decidedly un-bathroom-like vibe thanks to such touches as a freestanding display cabinet, a chandelier, walls that are painted rather than tiled, and inset floor tile that evokes an area rug. The shower windows, once blocked by a closet, now fill the space with sunlight.

Giving the toilet its own room adds to the un-bathroom vibe and makes the main bath more pleasant. Separating the toilet and the vanities from the tub and shower also makes the space more versatile for two users.

Closet

Master bedroom

Master bathroom

Makeup station

Toilet room

Vanities

LIGHTING

The bathroom is a multi-functional space. It's important to select

light sources and locations that will put the right type of light in the right spots

and help make the room a delightful place to be.

As we demand more function and beauty from our bathrooms, lighting has come to play an increasingly important and complex role. Rarely does a single ceiling fixture cut it these days—you need at least two types of lighting—task lighting and ambient lighting—and, sometimes, accent lighting.

Lighting Designer Peter Romaniello of Conceptual Lighting in South Windsor, Connecticut, advises starting with the task lighting and building from there. After all, he notes, "Bathrooms are very task-oriented rooms." Task lighting typically means lighting around the vanity mirror for shaving and applying makeup, lighting the shower and or tub, and lighting over the toilet if it is in its own enclosure. From there you can think about ambient lighting—the general lighting that fills the room, including natural light. If appropriate, you might add accent lighting to highlight a piece of art or other feature in the bathroom.

top right • You can purchase vanity mirrors that have LEDs incorporated into their surface, providing excellent light and a slick look.

right • Lights to each side of the vanity mirrors provide necessary task lighting, while ample ambient light comes from an overhead fixture and a window.

facing page top • Here's another way to light a vanity mirror from the sides. What could be better for anticipating a day at the beach than grooming in natural sunlight while enjoying a view out the window?

facing page bottom left • With four bulbs, this stylish ceiling fixture complements the bathroom's contemporary style while providing all the light this small powder room needs.

facing page bottom right • The light source hidden behind this vanity mirror adds drama and emphasizes the textured tile. While LEDs are fine positioned against textured tiles, they can reflect annoying pinpoints of light when used with smooth, shiny tiles.

Ambient Lighting

Ambient lighting fills the bathroom with an unfocused light that augments task lighting, or it can be used alone for such tasks as hanging fresh towels or getting a glass of water. Ambient light sources can be as simple as an inexpensive recessed overhead light or a window during the day.

One popular option is to install a fan/light combination in the ceiling. If you do this, be sure to get a fixture that can be wired to switch the fan and light separately. This is because you want to turn the light off immediately after leaving the bath while the fan should run after you leave—it can take 15 or 20 minutes to exhaust all the moisture after a shower.

With the right exposure, a window can provide all the ambient light you need during the day. If you are installing new windows, it pays to buy high-efficiency models that are rated for your climate and sun exposure. Talk with your architect about appropriate choices or visit the website of the National Fenestration Rating Council (nfrc.org) for a thorough explanation of window characteristics. Skylights, both the typical glass panels and the tube type that draws in light from the roof and diffuses it into the room, are another way to introduce ambient daylight.

Very small bathrooms may not need separate fixtures for ambient lighting and task or accent lighting. In a small powder room, for example, a pair of fixtures flanking a mirror over the sink may provide plenty of light for the whole room without a ceiling fixture.

Although ambient lighting can be simple, it doesn't have to be. As mentioned, LED strips can be installed behind a molding near the ceiling. In a large bathroom with a high ceiling, contemporary low-voltage lights on cables or even a chandelier might add style. One note of caution if you're considering a chandelier or other hanging fixture: The National Electrical Code prohibits hanging fixtures that can be reached from a bathtub.

Here's a useful tip: If you'll be ganging switches together in one box, put the switch for the ambient lighting closest to the door. This way you won't be fumbling for the switch in the middle of the night.

A wall-mounted fixture is just the ticket for reading in the bathtub. Here, the shape of a shelving nook was designed to leave wall space for a fixture.

Using the Color Rendering Index

Bathrooms are typically where women apply makeup and both women and men often check that clothing and accessories are a good color match, so it's important that light sources make colors look natural. The color rendering index (CRI) rates how realistic color will look under a given light source with 100 being the most natural.

CRI wasn't an issue in the days when most of us used only incandescent lights—they produce a warm, friendly light and have a CRI of 100. Halogen bulbs also have a CRI of 100. Fluorescent light tends to add a cooler, slightly bluish tone; good-quality compact fluorescent lights (CFLs) have a CRI of 80, which is better than the old fluorescent tubes. Most LEDS have CRIs of around 90, although some are as high as 97.

above • Light sources behind panels at the ceiling and above the tub provide ambient lighting in this bathroom while not detracting from the minimalist design.

left • Two recessed fixtures rather than one in a shower enclosure ensure you have light in front and in back of you. They also do a better job of highlighting the tile and water fixtures.

Adding Natural Light without Sacrificing Privacy

No artificial light can match the pleasure we get from natural sunlight streaming through a window. However, a typical clear glass window often won't work in a bathroom because of privacy concerns. Fortunately, there's usually a solution—two obvious ones are skylights and windows with opaque glass. Here are examples of four architecturally interesting alternatives.

1. This shade is designed to pull up instead of down, so you can leave the top part of the window uncovered while covering the bottom part for privacy. **2.** Glass block has long been used as a way to let light into a private space. It has excellent insulation value. **3.** Stained and opaque glass offers an opportunity to admit daylight while preserving privacy and adding art to the bathroom. **4.** All the windows in this bathroom are set high enough to provide privacy. Plus, you still get a view when standing. Lighting hidden behind the ceiling molding provides ambient light at night. Shiny tile reflects and spreads ambient light.

Accent Lighting

Accent lighting draws your eye to a feature you want to highlight without drawing attention to the light source itself. The feature might be a shelf where glass jars of toiletries are displayed, a piece of art, textured tile, even a sculptural soaking tub. There's no functional need for accent lighting—it's just there to add visual interest to the bathroom. Your design may call for no accent lighting at all.

Two categories of accent lighting are typically used in bathrooms: One type highlights objects with recessed pinpoints or spotlights. In the other type, more dispersed light washes or grazes a surface—usually a wall—to emphasize its texture.

Pinpoint trim kits are available for regular recessed fixtures—often the kits allow you to adjust the angle of the light. Low-voltage halogen lights have long been used for pinpoint lighting because they produce a warm, focused light. Nowadays, LEDs, which produce a similar light but last longer and don't require a low-voltage transformer, are used more often.

Some recessed lights do have an Achilles' heel. In bathrooms located directly below an unconditioned attic or right below an insulated roof, recessed fixtures can let warm, moist air escape. Not only is this a waste of energy, but it also can cause condensation problems. Make sure the fixtures are airtight and rated for contact with insulation. One way to ensure there will be no air leaks is to install recessed lighting in soffits rather than cutting holes in the bathroom ceiling.

To emphasize the texture of a wall, light needs to rake against that wall. This can be accomplished with a strip of LEDs mounted just behind a vanity mirror to light the walls around the mirror. You can also mount LED strips behind molding or a soffit at the top of the wall. But instead of allowing the light to escape upward to reflect off the ceiling as ambient light, the light is directed downward to wash the wall. Wall sconces can also be used to create light that accents wall texture.

Ambient lighting sources can be hidden or they can be celebrated, like this striking orange fixture that is a focal point against otherwise muted tones.

left · Each of these ceiling fixtures contains three lights that can be independently aimed for different purposes, including accenting a piece of art and providing more light in front of the tub.

bottom left · During the day, a round skylight provides most of the ambient light with a little help from round portholes in the wall behind the tub. At night, a strip of lights hidden in a recess near the ceiling does the job.

bottom right · Two pinpoint accent lights over a painting in a niche create a focal point in this bathroom.

Choosing a Light Source

LED technology has been improving while the price has been steadily dropping. As a result, LEDs have become the lighting source of choice in most applications—especially in the bathroom and kitchen because LED is more versatile than other light sources.

There are screw-in LED bulbs for use in standard lamp bases. These still cost more to buy than the incandescent bulbs they replace, but they use at least 75% less energy and they last 25 times longer. There is also LED "tape" or "rope" that consists of strips of tiny individual LEDs—spaced an inch or so apart—that can be cut to any length you need. These can be used under wall-mounted vanities as a dramatic nightlight. Or they can be mounted around the vanity mirror and capped with a diffuser strip—a particularly useful solution if space around the mirror is limited.

Bathroom designers are using LED strips to other dramatic effects. They can be installed behind a cove molding to bounce off the ceiling as ambient lighting. Also, lighting designer Romaniello notes, many designers are incorporating recesses or niches in shower walls with LED strips tucked into the top of the recess. The light rakes down the shower wall, a particularly interesting effect if you want to emphasize textured tiles. LEDs can be made to emit any color, so clusters of different-colored LEDs can be connected to a controller that allows you to change the color of the lighting to suit your mood.

While LEDs open new design opportunities, they also open up more possibilities for mistakes, says Romaniello. LEDs are available in various color temperatures, so if you are using LEDs in combination with other light sources, it's important to match the color temperatures. For example, incandescent bulbs produce a warm light that is 2,700 on the Kelvin scale, so you'll want LEDs with a 2,700 Kelvin rating.

A LED strip runs atop the full length of this mirror over a long countertop.

Dimming CFLs and LEDs

Installing dimmers in the bathroom is a good idea—dimmed lights can help to create atmosphere. Besides, no one wants to get blasted with full light when getting up to use a bathroom in the middle of the light.

When incandescent bulbs were the only choice, dimming was simple—all dimmers worked with all bulbs. But those "legacy" dimmers won't work well, if at all, with the much lower wattages of CFLs and LEDs. If you want to dim CFLs and LEDs, you'll need to install dimmers designed to work with them. Fortunately, most new dimmers designed for LEDs will work with CFLs as well. Not all CFLs and LEDs are dimmable and not all dimmable bulbs work with all dimmers. Each dimmer model will come with a list of bulbs that will work with it. So purchase your dimmers and then make sure you get compatible bulbs.

An accent light rakes these light blue tiles to highlight their unusual bubbly texture.

Also, LEDs produce bright, directed light. Diffusers will spread the light, but if you put an LED bulb in a fixture that was designed for an incandescent bulb, the light might glare through the diffuser cover. An increasingly popular application is to install LED strips behind the perimeter of a mirror. A word of caution: If wall tiles are shiny they can reflect the individual diodes as annoying bright spots. An alternative is to purchase a vanity mirror with frosted LED lights incorporated into the surface, available from a company such as Electric Mirror®. You can even find mirrors that let you adjust the light temperature so you can approximate how makeup will look in different situations, such as outdoors or at an evening event.

Because of the advances and price drops for LEDs, Romaniello sees no reason to use traditional incandescent or halogen lighting for any purpose other than in certain traditional-looking decorative fixtures. These are fixtures designed to be a part of the décor as much as a light source, and some would not look right with an LED bulb.

Another reason to include an incandescent or halogen fixture in your lighting scheme is if you plan to dim a light to use as a nightlight. Some LEDs can be dimmed, but they won't dim enough to use as a nightlight.

CFL bulbs are essentially obsolete, Romaniello believes. They last longer than incandescent or halogen bulbs but not nearly as long as LEDs. They are cheaper than LEDs, but the price gap is closing. Plus, most people don't like the squiggly shape of CFLs, they take up to a minute to reach full power, and they contain poisonous mercury.

If you have a generous budget and are designing a large complex bathroom, it might be worth your while to hire a lighting designer. At least, do your homework online and then visit a local lighting store for advice. Well-established retailers usually have knowledgeable staff.

Hidden accent lights wash the wall behind this vanity, emphasizing its texture.

Fixtures at both sides of vanities provide ideal task lighting.

A simple old-style ceiling fixture provides ambient light for this homey bathroom.

Light Sources

LEDS
$-$$$

- Increasingly popular as their price comes down and they become more widely available for residential use.
- Although still more expensive to purchase, extremely long life—more than 50,000 hours—plus much lower energy use can make LEDs the least expensive option in the long term.
- Available in different color temperatures from warm to white.
- More light output than CFLs.

INCANDESCENT BULBS
$

- Inexpensive.
- Produce a pleasing, warm light.
- Don't last as long as several others kinds of lamps.
- Inefficient—more than 90% of the energy they use produces heat, not light.

FLUORESCENT LAMPS
$-$$

- Use less energy to produce the same amount of light as an incandescent bulb.
- Much longer life than incandescent but much shorter than LED.

- CFLs screw into standard lamp bases, but squiggly shape may be displeasing if visible.
- Some types can be dimmed, and lamps with a high color-rendering index make people and objects look natural.
- Contain small amount of mercury and should not be thrown away with household trash.
- Have been falling out of favor as LED lighting improves and drops in price.

HALOGEN BULBS
$-$$

- Some halogen bulbs, including Halogena® bulbs from Philips, offer better energy efficiency than incandescent bulbs but not as much as CFLs or LEDs.
- Halogens emit a bright, white light and look very similar to standard incandescent bulbs.
- Unlike CFLs, they reach their full lighting potential as soon as they are turned on, and they contain no mercury.
- They have high operating temperatures, very good color rendering, and last longer than incandescent bulbs.
- Low-voltage halogens use transformers to step down 120-volt household current to 12 volts. The halogen bulbs used in low-voltage fixtures make them a good choice for accent lighting.
- Low-voltage cable and monorail fixtures are stylish and contemporary but usually pricey.

Lighting Fixture as Design Element

Some fixtures are hidden—their sole purpose is to cast light. Others are quite handsome and are chosen to enhance the bathroom's décor as well. Selecting the right combination of fixtures and combining their artificial light with natural daylight will make any bathroom more attractive and functional.

1. A sculptural fixture draws the eye and bursts with ambient light. A small window strategically placed at the head of the tub provides a view, ventilation, and some light without compromising privacy. 2. Traditional-looking fixtures blend with traditional architecture to create a very up-to-date lighting design. Globes flanking the mirrors provide task lighting, while ambient lighting comes from a matching pendant over the tub augmented by recessed ceiling lights and a bank of windows beside the tub. 3. To prevent unwanted shadows, vanity mirrors are best lit from the sides. However, if that's not possible, using fixtures with multiple bulbs over the mirror won't create the harsh shadows that a single light source would. 4. When privacy isn't an issue, use large windows to bring the outdoors in. Here, large mirrors reflect the outdoor scene, reflect more light into the room, and make the room seem bigger. Hanging fixtures float like bubbles and provide ambient light at night.

A Few Square Inches Make All the Difference

It started out with a very practical concern: The sink in this powder room needed to be recessed a few inches to allow 21 in. of clear access to the front of the toilet, as required by code. But that little recess opened up several creative design opportunities.

"It actually added some dimension to a powder room that otherwise could have been pretty boring," said Sophie Piesse, the architect who designed the powder room and the house in Carrboro, North Carolina.

The first step in turning a contingency into a striking feature was to add a long mirror surrounded by a cascade of thin green tiles that accentuate the height of the nook. The horizontal grout joints are randomly placed, while the vertical joints are aligned, adding to the perceived height. The sink is set over a small cabinet that hides the plumbing and offers just enough storage for a couple rolls of toilet paper.

Bathroom mirrors are best lit from the sides to prevent unpleasant shadows. Not long ago, that would have been tricky to do in such a tight nook. The answer was LED tape lighting: A narrow aluminum channel that can be cut to any length is attached to the wall and a tape embedded with tiny light-emitting diodes is cut to fit in the channel. A lens is fitted in place to diffuse the bright light so it fills the room.

"The owners had bought a pendant light to hang over the sink, but by the time all the details were finished they didn't use it," Piesse said. "There's plenty of light from the LEDs."

The advantage of this little niche extends beyond the bathroom to the mudroom on the other side of the wall, where the jog in the wall allowed a closet to be deeper than the bench and cubbies next to it.

"Sometimes," Piesse said, "a few square inches can make all the difference."

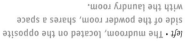

facing page • The jog in the powder room wall, slight as it is, allows the extra depth needed in the mudroom for the closet to the left of the bench and cubbies.

left • The mudroom, located on the opposite side of the powder room, shares a space with the laundry room.

left and above • This narrow powder room would have been pretty boring were it not for the slight bumpout around the sink. The niche is emphasized by green tile around the mirror and provides just enough surface to mount LED strip lighting that extends to the ceiling.

FIXTURES

When it comes to fixtures, form really does follow function.

You'll want to choose good-quality fixtures in a configuration that best suits

your needs and a style that best suits your taste.

Sinks

All sinks are pretty much equally functional and durable. Whether you swing by your local home center for a plain white $40 vitreous china model or pour $4,000 into a concrete sink designed and made to your specs, all sinks hold and drain water and they'll all last for decades.

And yet sinks, or lavatories as the pros call them, are available in a seemingly endless array of styles, and there is plenty to consider when choosing one for your new bathroom. For starters, how will it be used? You'll choose a different sink for a heavily used kids' bathroom than you would for an elegant powder room. With some careful thought, you'll be sure to find a model that perfectly suits your needs, taste, space, and budget. All those models fall into eight basic types, and picking one of these types is a good place to start in narrowing down your choice. Here's a look at the pros and cons of each type.

SELF-RIMMING

In this configuration, also called top-mounted or drop-in, the sink fits into a hole in the vanity with part of the sink overlapping the countertop. The least expensive sinks are top-mounted, but not all top-mounted sinks are inexpensive. At the top end, you'll find sinks that are handmade from exotic materials, including metals such as copper, and models that were hand-painted before being fired.

Because the rim covers the hole, top-mounted sinks are relatively easy to install and can be mounted on whatever countertop material you choose. For example, a rough hole can be cut into plastic-laminate-covered plywood. Self-rimming sinks work well with tile counters, again because the rim will cover rough tile cuts. These sinks often include holes for mounting the faucet rather than using separate faucet holes in the countertop.

The one disadvantage of top-mounted sinks is that when cleaning you can't wipe the counters directly into the sink, and the caulked joint between the rim and counter can collect dirt.

above • A basic inexpensive self-rimming sink can be the perfect choice for a family bathroom. It's easily installed in a standard vanity and will perform as well and as long as any sink you can buy. The oval shape of this model makes efficient use of counter space while allowing plenty of room for hand washing.

left · High sides give this self-rimming sink a bolder look that fits in with the geometric wallpaper and complements the solid wood countertop.

facing page bottom · The old-timey look of this trough-style self-rimming sink fits right in with the cottage architecture of this home. It accommodates two faucets into an integrated backsplash that's easy to keep clean.

An elegant self-rimming sink of hammered copper is set into a smooth copper countertop, enhancing the warm natural tones of the wood cabinetry and the tiles around the mirror.

UNDERMOUNT

These sinks are mounted under the hole in the countertop for a more streamlined look that emphasizes the design of the faucets and countertop rather than the sink itself. The faucet, of course, is mounted in the countertop. Undermount sinks are a good choice for a heavily used family bathroom because the uninterrupted counter surfaces are easily wiped clean.

Undermount sinks come in a wide range of styles and prices. Most are attached under the sink with clips, but they can also be glued directly to the bottom of the counter. You can also buy countertops with sinks that are premounted underneath, available at home centers. Keep in mind that if the sink is permanently glued to the counter, you won't be able to change one without the other.

Because the edge of the counter hole is exposed, undermount sinks are best used with solid waterproof materials, including stone and concrete, or monolithic man-made materials, such as cast polymer.

top right • A simple white undermount sink takes a backseat to an antiqued copper faucet and lovely marble countertop and backsplash.

bottom right • Careful tile setting around the perimeter of the undermounted sinks creates a waterproof countertop. The result is a striking juxtaposition of rigidly geometric counter, walls, and floors with a "live edge" walnut apron.

A rectangular under-mounted sink echoes the shape of a massive concrete countertop.

A glass sink in light blue undermounted in an aquamarine countertop creates a striking ultramodern effect.

A bronze sink embossed with lilies is at the high end of under-mounted models.

INTEGRAL

Here, the sink and the countertop are made from one piece of the same material—ceramic and cast polymer are common because they are easily molded. For considerably more money you can get beautiful integral sinks and countertops in materials such as hammered copper or stone. Because they are seamless, integral sinks are easy to keep clean.

Sink and counter combos come in various shapes, sizes, and styles. Some have sinks that slope in from one side. Sometimes the counters on both sides are sloped to allow splashed water to run into the sink, and some counters have raised edges to catch water.

In this eclectic bathroom, the vanity of rough-sawn planks contrasts with the smooth lines of the gradually sloping integral sink.

Some integrated sinks are designed to accommodate two faucets. Here, a simple removable cross-shelf made to match the shelf under the mirror adds an accent while providing a little more space to set down makeup or a toothbrush.

An oval integrated sink complements the clean lines of this vanity. The graceful faucet set matches the pulls on the vanity.

An old-style wall-mounted sink with integrated backsplash adds to the old-fashioned charm of this traditional white bathroom. You can tell that the sink and faucets are not antiques because the hot and cold water mix into one spigot instead of the separate spigots on old faucets.

This sleek wall-mounted sink features a wooden surround that hides the plumbing and matches the unique shelf/mirrored medicine cabinet above.

WALL-MOUNTED

These sinks are attached to the wall with no support below. Because they don't interrupt the floor space, wall-mounted models make the room seem more spacious, but of course they also eliminate the storage that a vanity would provide. As a result, they are a good choice for a small powder room where storage isn't important.

If your bath remodel plan doesn't call for stripping the walls down to the studs, be aware that you'll need to open the wall behind the new sink to install support blocks between the studs.

With wall-mounted sinks you'll see the drain trap and two supply pipes below, although some models sold as semi-pedestals include a shroud that covers the plumbing. With some models, the water supply lines can be tucked out of sight. You can also turn the drain into a design feature by purchasing a drain trap in a style and finish to match the faucet.

Most wall-mounted sinks are accessible to someone in a wheelchair. You can buy soft plumbing covers to protect wheelchair users' legs.

Here, a sink in the form of a simple bowl is fitted to the wall with brackets. Its faucets are mounted through the wallpaper. The wallpaper behind the sink might not hold up to daily family use, so this elegance is best reserved for a lightly used powder room.

Wall-mounting a sink in a corner is an efficient way to make use of space in a tight powder room. Sinks mounted on a single wall must be designed for that type of mounting with solid blocking behind the wall. Corner sinks can be supported by cleats mounted to the surface of both walls.

PEDESTAL

A traditional look may come to mind when you think of pedestal sinks, and many models do fit right in with Victorian moldings and beadboard wainscoting. But there are plenty of contemporary-looking models available, too. Pedestal sinks are often part of a suite of fixtures, so you can buy a toilet and tub with matching details if you like. They come in various sizes, including models designed to tuck into the corner of a powder room.

In common with wall-mounted and two-legged console sinks, pedestal sinks require blocking inside the wall and they preclude storage below. Most of the drainpipe is hidden in the pedestal. The drainpipe, along with the water supply lines, will be visible where they enter the wall, but they'll be unobtrusively obscured by the basin and pedestal.

As is the case here, pedestal sinks are often part of a suite of fixtures that includes the sink, toilet, and tub. This suite is a perfect match for the subway-tile walls and white and grey floor tiles.

It's not really a pedestal sink—maybe more like a standing vessel sink. In any case, its orange interior and white exterior and its shape seem custom made for this strikingly out-of-the-box bathroom.

This sink brings to mind a bowl on a stand, which is essentially what you have when you reduce a pedestal sink to its elemental forms.

Pedestal sinks don't offer much counter surface. Here, flanking cabinets provide a handsome solution.

This bathroom is a study in contrasting textures, including the free-form vessel sink that's smooth on the inside and textured on the outside.

VESSEL

Vessel sinks are essentially bowls mounted atop a counter or sometimes recessed partway into the counter. Most vessels are round bowls, but because they don't have to fit a counter hole, they can be any shape, from rectangular to swooping sculptural forms, and they can be made of any material that will hold water, from glass to brushed nickel to ceramic pottery. Many are essentially functional works of art that make a great focal point for a powder room that will get light use, mostly by guests— we're not talking about the downstairs half-bath that the kids use daily.

Vessel sinks do have some special considerations. Keep in mind that when mounted on a standard-height vanity, a vessel sink will be higher than a standard recessed sink. Depending on the height of the vessel this may be fine, but it's something to be aware of when designing your bathroom. Also, vanities of standard depth may be too shallow for some vessels.

Because of their high sides, vessel sinks are most often coupled with wall-mounted faucets, which limits your faucet choices. Vessel sinks are more prone to splashing than other types, so the faucets need to be carefully selected and positioned to minimize this problem. Because the outside of the sink is exposed, there is more surface to clean, and some of that surface may not be easily accessible.

left · Perfect for a vacation cabin bathroom, this vessel sink set on an antique chest evokes the pre-plumbing era when folks poured water into a basin to wash up.

top • Vessel sinks of translucent glass are available in any color you can imagine. This one works nicely with the glass-tile backsplash in green and two hues of blue.

above • A delicate-looking glass vessel in the shape of an inverted pyramid is sharp both in appearance and in its top edges—not a good choice for a bathroom that kids will use.

right • The undulating rim on this otherwise simple vessel sink evokes the motion of the river visible through the porthole windows. Its oval shape is as practical for hand washing as it is beautiful.

SEMI-RECESSED

If you need a full-depth sink but don't want to devote the space required for a full-depth vanity cabinet, a semi-recessed sink may be the ticket to a successful design. Some of these sinks allow you to use a cabinet as shallow as 12 in. These sinks are a bit more difficult to install because you have to cut out the front of the cabinet as well as the countertop to slide the sink into place. A semi-recessed sink overlaps the countertop like a top-mounted sink.

above • Semi-recessed sinks overlap the front of the vanity—a handsome way to save vanity depth while enjoying a full-size sink.

left • The counter is recessed into the wall while the sink and counter are semi-recessed into a shallow vanity. The result creates an attractive nook while using very little floor space.

A Wet Room Optimizes Available Space

There wasn't much floor space to devote to the main bath in this modestly sized, newly built bungalow in a Minneapolis neighborhood. A sloped roof further encroached on the available space.

The solution by architect Paul Hannan of SALA Architects was to skillfully blend the modern concept of a "wet room" with the style of a 1930s bungalow. In this wet room, the shower and bathtub share the space under the sloped portion of the ceiling. A fixed glass panel spans half the opening to the wet room, containing the shower spray without the need for a space-eating door.

The tub is tucked under the lowest part of the ceiling, creating a cozy bathing spot in an area that's too low to stand. There is no curb at the entrance to the wet room. Instead, the floor is sloped to a drain in the shower space so the octagonal floor tiles can flow uninterrupted from the wet room to the rest of the bathroom. This makes the bathroom seem more expansive. It also eliminates a potential tripping hazard and makes the shower accessible to a wheelchair.

Two windows flood the bathroom with light. To create a handsome exterior façade, the windows are identical from the outside. However, the window in the wet room is covered with a frosted glass panel that provides privacy and protects the wooden window from moisture damage.

The walls and ceiling of the wet room are covered with subway tile. The same tile is dropped down to become a wainscot in the rest of the bathroom. This defines the spaces and adds visual interest. It also costs less than completely covering the walls with tile. A band of small black tiles runs around the entire space, tying it together in another visual trick that makes the space seem larger than it is.

A floor-to-ceiling cabinet with drawers below maximizes storage. It's located strategically in a corner so it doesn't interrupt the openness of the space or block light from the windows. It's also set back from the vanity just enough to keep from crowding the door.

A narrow band of black tiles sweeps through the entire bathroom space, helping to make it seem bigger than it is. Notice that the faucet is mounted on the outside corner of the tub, making the controls more accessible under the sloped ceiling.

above · The plumbing fixtures are much sleeker than what you would find in a bungalow of the 1930s. But their chrome finish evokes the period just enough to make them fit right into this new bungalow.

left · The tub is nestled under the lowest part of the sloped roof, maximizing the use of space in the wet room. No curb interrupts the flow of floor tile, and a partial panel is all that's needed to prevent shower water from splashing into the rest of the bathroom.

Countertops

Countertops tend to be a focal point in the bathroom, so it's important to choose something that enhances the look of your design. Beyond appearance, the palette of materials at your disposal varies in cost and maintenance requirements. And if green building is important to you, there's a growing array of beautiful materials fabricated from recycled glass or even recycled paper.

As with everything else in the bathroom, think about how the countertop will be used. A visually striking slab of exotic wood might be the perfect choice for a small powder room counter—and perhaps not a budget buster since you'll only need a small piece. However, in a room that regularly fills with shower steam, a wood counter will need careful finish and maintenance.

In choosing and budgeting for countertops, keep in mind that the cost of the material itself is only part of the equation. Some materials, including plastic laminate and wood, can be cut on-site to fit. Most other materials require fabrication in a specialized shop or plant, adding to their cost unless you are buying a pre-made counter to fit a standard-size vanity.

below • As sailing enthusiasts, the homeowners are familiar with the beauty and rot-resistance of mahogany used in building boats, so they chose mahogany with a marine varnish for this countertop and vanity.

PLASTIC LAMINATE

Plastic laminate, often referred to by the brand name Formica®, is a very durable and cost-effective choice. However, its popularity has been decreasing. One reason may be that you can buy a solid-surface top with an undermount sink already attached, sometimes sold along with a matching vanity. Because these packages save installation labor, their cost can be competitive with having a plastic laminate countertop custom-fitted to a vanity. Of course, buying a premade package limits your choices of counter and vanity styles, colors, and sizes, while plastic laminate, and for that matter separately purchased solid-surface counters, are available in a wide range of colors and patterns.

SOLID SURFACE

Solid-surface counters are made of acrylic or polyester or a blend of the two plus a filler. The material is available in any color or pattern and doesn't stain. Because it can be molded into any shape, the material is used to make seamless one-piece sink/counters that are easy to clean.

Solid-surface material is available in virtually any pattern or design including the vivid blue used for this counter and the bench platform under the cushion in this bathroom.

STONE COMPOSITE

A variation on solid-surface material is stone composite, also called engineered stone or quartz. It's made of 90% quartz combined with acrylic and/or polyester. Like solid surface, engineered stone doesn't stain and is easy to clean. It's also made in many colors and patterns and can mimic various types of stone. Engineered stone generally costs a bit more than solid-surface material but less than natural stone. The quartz makes engineered stone harder than other solid-surface materials and gives it a more natural look.

A Technologically Advanced Countertop

Dekton®, made by Cosento, is a unique proprietary product that has been on the market just a few years. Like solid-surface material, it can be molded into various colors and textures. According to the company, it is made from "inorganic raw materials found in glass, porcelain, and natural quartz" that are combined under pressure—no resin is used. The product requires no sealing, and the company says it is very resistant to heat, stains, and scratches. However, Dekton is also one of the most expensive counter materials you can buy.

NATURAL STONE

There are many types of natural stone available—granite and marble are the most popular choices for vanity tops. These stones are porous and need to be sealed to prevent stains. Soapstone is more expensive but doesn't need to be sealed.

TILE

Tile is a versatile alternative that comes in an endless array of sizes, colors, and texture, so you can achieve a customized look. A tile countertop can match or complement tiles used elsewhere in the bathroom. Keep in mind that grout lines can collect grime, making tile countertops a bit harder to clean than smooth surfaces.

facing page top · The homeowners knew they wanted the durability of stone for this countertop. They selected green granite to complement the reddish Douglas fir trim and vanity.

facing page bottom · Semi-precious stone can be sliced and set in clear epoxy to create striking countertops. This one features slices of agate.

left · The subtle color variations of this pale tile counter and backsplash are a counterpoint to this bathroom's warm wood details.

GLASS

Glass is striking in appearance and very easy to clean. It's available colored or clear, textured or smooth. Its cost is comparable to granite. Several manufacturers make counters from recycled glass. In some cases, recycled glass is incorporated into a product that is indistinguishable from new glass. Other products combine chips of glass with a binder of either cement, which needs to be sealed, or resin, which doesn't.

PAPER

Paper is not the first material you'd think of for a countertop, but some manufacturers are combining recycled paper with phenolic resin to make durable, easy-to-clean countertops that cost about the same as granite. Phenolic resin has been around since the early 20th century when it was used to produce the Bakelite® in old black telephones and early radios.

CONCRETE

Another option is concrete, which can be cast into any texture or shape, including a counter with an integral sink, and tinted to the color of your choice. It can also be embedded with decorative stones, glass, tile, or any other hard, durable material you might desire. Concrete is always custom-cast for the job, which allows lots of design freedom but also makes it an expensive option. Some cabinets may need to be reinforced to support concrete countertops.

top right • This glass countertop mimics flowing water. Mounted against a glass wall, it blends with a beautiful wetlands view.

right • This colorful custom countertop is made of recycled glass set in epoxy resin. The homeowner selected the glass colors to be used and the percent of each to be included.

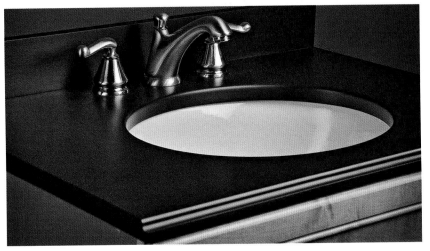

above • Taking full advantage of the versatility of concrete, this countertop includes a sloped sink and flows around a corner to become the tub surround.

left • This countertop from PaperStone® is made of recycled paper and phenolic resin. It is designed to mimic slate.

Faucets

Most faucets are designed to fit a specific configuration of holes in the countertop or sink, and, as mentioned, not every countertop material works with every sink so it's best to shop for all three at the same time. Self-rimming and pedestal sinks usually come with holes along the back edge for mounting the faucet, as do countertops that have integral sinks or are designed for undermount sinks. The number of holes and their spacing determines which faucets will fit. When it comes to hole configuration, all faucets fit into one of four categories:

Centersets. These faucets have the spout and two handles or a lever mounted on a shared base. They are compact, making them well suited to small sinks and tight spaces. Centersets fit into three closely spaced holes.

Widespread faucets. These have two handles and a spout that are each mounted directly to the counter or sink without a shared base. As the name implies, the holes are spaced farther apart than holes for centersets.

Monoblock faucets. In this configuration, everything is housed in a single component. Most often, a single handle is mounted atop the spout, but there also are models with a hot-water control handle mounted to the left of the spout and a cold-water handle to the right.

Wall-mounted faucets. These have the spout and one or two control handles mounted to the wall behind the sink. They are trickier to install because all the plumbing is in the wall, but they do make it easier to clean around the sink. Wall-mounted faucets are often paired with vessel sinks.

Other features. Beyond mounting type and number of control handles, there are a couple of other features to consider. Faucets are available with motion sensors to turn the water on and off so you don't need to touch handles. They also conserve water. Most faucets come with the familiar aerator that mixes air into the water, creating a whitish stream. You might prefer a faucet that comes with a laminar flow device that produces a clear stream. (You can also buy laminar flow devices separately to replace aerators.)

above · Monoblock faucets mount into a single hole to create a spare and elegant look.

right · Wall-mounted faucets pair well with vessel sinks. Many counter-mounted faucets are not tall enough for use with vessels.

facing page top · Centerset faucets have the spout and handles mounted on a single escutcheon, a compact configuration that is great for tight spaces.

facing page bottom · Widespread faucets fit into three holes, better for installation into countertops and larger sinks.

Think of your faucets as little machines that get used every day—the quality of construction really does matter. Look for faucets that are heavy, with parts that move smoothly with little play. Almost all modern bathroom faucets use one of two methods to start and stop the flow of water. The most durable and costly faucets have control handles that use two highly polished ceramic disks. One disk is fixed in place and slides over the other disk. The disks actually grind away deposits that can cause leaks. The other method is a cartridge that wears out more quickly but can be easily replaced.

Once you've decided which type of faucet makes sense for your bathroom, you can start thinking about aesthetics. You'll have no problem finding faucets, along with matching showerheads, that fit any style you've chosen for your bathroom, from traditional to ultramodern. Finish is part of the aesthetic choice. You'll find chrome, nickel, brass, copper, and bronze as well as bronze sold as "oil rubbed"—a chemically darkened finish designed to simulate aged bronze. Top-quality finishes are applied with a process called physical vapor deposition that creates a more wear-resistant surface than conventional electroplating.

top · The dark oil-rubbed bronze of the widespread sink faucets works with the cabinet pulls and other dark elements as a counterpoint to the warm natural wood tones in this bathroom.

right · This unusual modernistic faucet features a flat chrome spout with a flat control lever to the side.

top left • Nickel widespread faucets are among the details that contribute to the elegance of this bathroom.

above • This monoblock faucet with handle on top fits perfectly into the tight space allotted to this corner sink.

left • This gold wall-mounted faucet works with an integrated sloping sink that features a linear drain at the bottom. The result is a unique look that is easy to clean.

Toilets

If you want your toilet to light up and open when it sees you and then play music while it warms your seat and feet, and if you want it to offer a bidet and then close and clean itself after you leave, the Kohler Numi® is one model that will give you all of that—for around $7,000.

More likely, you're in the market for a simpler device—one that flushes reliably, uses water efficiently, and is easy to clean. If so, you have plenty of good choices. Since a 1995 federal regulation requiring toilets to use no more than 1.6 gallons of water, manufacturers have been developing efficient designs that work very well while meeting the standard. California has an even tougher law—toilets sold in that state are required to use no more than 1.28 gallons per flush.

Still, there are a number of important decisions involved in choosing a toilet that suits your needs and design, so don't leave the decision to the plumber. After all, the humble toilet was the original reason for the "water closet." Start by asking yourself some questions:

Round or elongated bowl? This decision is informed by how much room you have allocated for the toilet. Round bowls are more compact and generally less expensive, but many people find that elongated bowls are more comfortable. There are also compact elongated bowls, a compromise that requires a little more space than round bowls.

How high? Toilet seats typically are 14 in. or 15 in. high—a height many people find makes the toilet most comfortable to use. You'll also find toilets that are 16 in. or 17 in. high. Often sold as "comfort height," these models are helpful to people with mobility issues or who need to transfer from a wheelchair.

One piece or two? Most toilets are in two pieces—the tank is bolted to the bowl during installation. One-piece toilets are easier to clean because there is no seam between the bowl and tank where grime can collect. They are, however, more expensive and a bit harder to install.

Dubbed "comfort height" by manufacturers, taller toilets were originally designed to meet requirements of the Americans with Disabilities Act. Shown here is a two-piece unit with a round bowl.

Many people find elongated bowls more comfortable than round bowls. This dual-flush model has two flush buttons atop the tank.

At about $3,000, this sleek Veil® toilet from Kohler features a heated seat, bidet wand with adjustable water temperature and pressure, warm-air drying, UV self-cleaning functionality, and a LED nightlight in the bowl.

Does a dual-flush make sense? If saving water is important, consider getting a dual-flush toilet that gives you the option of a full-flush, typically 1.28 gallons, when you need it, or a partial flush, usually just 1.1 gallons, when that's all you need. Many of these toilets work with two buttons on the top; some newer models have two levers on the side—a longer partial-flush lever with a shorter full-flush lever stacked on top of it.

What about wall-hung? You might associate wall-hung toilets with public restroom stalls, but the availability of sleek upscale designs has made them increasingly popular for residential bathrooms. With the water tank hidden in the wall, wall-hung toilets take up significantly less space, and they are easy to keep clean. Another advantage is that you can mount them at whatever height suits you best. Most residential wall-hung units are dual flush. The flush buttons are mounted on a large plate in the wall behind the toilet. The plate can be removed to gain access to the tank if the flush mechanism needs to be repaired or replaced.

Wall-mounted tanks come mounted on a frame designed to fit between wall studs, so installation is not too difficult during new construction or if your bathroom renovation calls for gutting the walls. Otherwise, a wall-hung unit probably isn't a practical choice, especially if you are replacing a floor toilet.

Wall-hung toilets fit right in to sleek modern design. Like most wall-hung units, this one is dual flush: The big button on the wall plate is for a full flush, while the small button is for a partial flush.

Many dual-flush toilets have two flush levers—one for a partial flush and one for a full flush. On this Kohler model, the partial-flush lever is green as an intuitive visual reminder.

CHOOSING A LOOK

Toilets come in a variety of traditional and contemporary styles and, as mentioned, you might want to choose the toilet, tub, and perhaps a pedestal sink from a manufacturer's matching "collection." You can find toilets in a variety of colors, but it's not surprising that white has long been by far the most popular choice. White does make sense for a sanitary device, but beyond that white never goes out of style and doesn't clash with any other colors. Choose a white toilet with a matching white tub, and you won't have to worry if you decide to paint the bathroom a different color five years down the road.

With most traditional-looking toilets, the base casting clearly shows the curved water passageway molded into the side. This creates more nooks and crannies to complicate cleaning. More contemporary-looking toilets have a smooth, seamless base that is easier to clean.

If you really want a nod to the past, you can still buy a toilet with a tank that gets mounted high on the wall and operates with a long pull-chain. Now simply a style choice, this was originally a design of necessity—early toilets needed the water to gain velocity to do its job.

SPECIAL-PURPOSE TOILETS

The typical household toilet is a simple device that works well in most situations: When you pull the handle, a stopper lifts, allowing water to pour from the tank into the bowl. When you release the handle, the stopper falls back into place and the tank fills with water until a float reaches a point where it turns off the water. The mechanism lasts for years and when it does wear out, you can buy a replacement kit at any hardware store for about $12. There are, however, two other types of flush mechanisms designed for special situations:

MACERATING TOILETS
Are you finishing the basement and adding a toilet? A standard flushing mechanism relies on gravity and won't work if the fixture outlet is below the waste line. For this situation, you'll need a macerating toilet. Instead of sucking waste down through a waste pipe in the floor, these toilets have an outlet

This traditionally styled toilet reveals the curved water passageway in the base.

Low-profile one-piece toilets have a contemporary look, are less conspicuous, and are easy to clean.

at the back that connects to a macerating unit that grinds up the waste and pumps it up to the waste line. These units require an electrical outlet. If you have a shower and sink in the bathroom, they can be connected to the same unit. Some macerators can pump waste up to 16 ft. vertically and 150 ft. horizontally, making it possible to build a bathroom in areas that would otherwise be inaccessible.

PRESSURE-ASSIST TOILETS

That loud whoosh you hear when you flush the toilet at a public rest stop is probably because the toilet uses the pressure in the water-supply lines to store a charge of compressed air in the tank. When you pull the handle the pressure is released, creating a powerful flush designed to eliminate waste quickly and clean the bowl thoroughly. The next user doesn't have to wait for a tank to fill before flushing again.

These fixtures can make sense in high-traffic bathrooms at home under special circumstances. They use less water than conventional toilets, for example. But, in general, modern toilets are efficient enough, and this feature adds unnecessary complexity and cost.

COMPOSTING TOILETS

Today's good-quality composting toilets look pretty much like other toilets and do an effective job of removing odor. They do have the advantage of using no water while producing compost for the garden. However, the type you would want in your home is pretty expensive. Most people would consider a composting toilet only if there is no sewage or septic hookup available. Many models require an electrical hookup to run a thermostatically controlled fan and heater that quickly biodegrade the waste, reducing its mass by more than 90% and turning it into humus you empty every couple of months. There are also nonelectric models. In both types, positive airflow controls odor.

left • This serene bathroom is designed for easy cleaning. Although the toilet is a two-piece unit, the tank transitions smoothly into the base, which has no nooks or crannies to collect grime. It's easy to clean under the raised vanity, and continuing the tile up the walls without a baseboard eliminates another grime-collection spot.

Bidets and Urinals

BIDETS

The bidet was invented in France in the late 17th century. Bidets have long been popular in Europe, and they have become extremely popular in Japan, where more than 80% of all households have one. Yet they continue to be the exception rather than the rule in American homes.

Traditionally, a bidet is a low basin where you can wash up after using the toilet. They're useful for both men and women, and they're available in a variety of styles from most major fixture manufacturers. Bidets require hot- and cold-water supplies. Just like a sink, the faucets are sold separately from the basin, and you need to match the basin's hole configuration.

Bidets are typically installed next to a toilet. If you don't have the floor space for a separate bidet, you can buy a toilet that includes a bidet spray. Another option is to add a bidet seat to a standard toilet.

top right • You can purchase a seat that adds bidet functionality to a standard toilet.

right • Bidets are typically paired with a toilet in the same style. Faucets are sold separately.

top • A urinal fits right in with the very masculine design of this bathroom.

left • This whimsical (and perhaps a bit musical in use) urinal from Philip Watts Design is a galvanized bucket on a wooden base.

URINALS

Including a urinal in your bathroom plan might not have occurred to you. Of course, you don't really *need* one, but there are some advantages worth considering. On the practical side, urinals use less water than even a dual-flush toilet; in fact, there are waterless models available. And urinals are easier to keep clean than a toilet.

Urinals designed for homes are a lot classier looking than what you'll find in a public restroom. Philip Watts Design, Villeroy & Boch®, Duravit®, and Kohler offer contemporary designs. Men simply find urinals more convenient than a toilet, and they can be a great addition to a masculine bathroom.

Blending New with the Best of the Old

Smart renovation is as much about what you keep as it is about what you replace. That's especially true when you are bringing new life to an architecturally significant period home like this grand 1920s house in Philadelphia.

The large master bathroom already had beautiful marble tile on the walls and floors. A few broken tiles needed replacement, but beyond that, a thorough cleaning was all that was needed to bring these surfaces back to their original glory. Also existing was an old claw-foot bathtub with an unusual, if somewhat impractical, wooden rim and brass faucet. The toilet and the glass-enclosed shower, dating from a previous remodeling in the 1980s or '90s, were also retained, although the showerhead and valve were replaced.

And oh yes, there was the fireplace—a luxury as unexpected in a bathroom of the '20s as it is today. This one was no longer safe for burning wood, so architect Debbie Woodward of Kreiger + Associates had a gas-fired insert installed.

What the space lacked was storage. To solve this problem, the old pedestal sink was replaced by a custom-made vanity with cabinets. A couple of pre-existing factors made this piece challenging to design. First, to avoid blocking the heater, the space under the counter had to be left open, forgoing the storage space typical in an enclosed cabinet. Besides that practical consideration, the homeowner wanted the more period, freestanding look of a piece with legs. To provide storage, a pair of cabinets flanks the sink. Below each cabinet are a drawer and two open shelves.

The second challenge was to avoid blocking light from a window next to the vanity. To solve this problem, the sides of the cabinets, as well as the fronts, have glass panels.

The vanity was the one major addition to this bathroom. The cabinets have glass sides as well as fronts to allow light from the window to pass through. The open space below allows heat from the baseboard under the window to flow into the room. The new wall-mounted light fixtures continue the period motif.

The glass shower from an earlier renovation received a new period-style showerhead and valve but was not otherwise changed.

The fireplace that once burned wood was fitted with a gas-burning inset.

The soaking tub with its unusual wooden rim was left just as it was.

The homeowner loves this old mirror, so the vanity was designed to incorporate it. The vanity top of Bardiglio marble picks up on the dark streaks in the wall marble, complementing the existing marble without trying to match it.